OFF-GRID SOLAR POWER

FOR BEGINNERS

Design, Install, And Maintain Off-Grid Solar Systems

For Homes And Rvs

Patrick C. Lau

Copyright © 2024 by Rosy Gilbert

All rights reserved. No part of this book may be reproduced, distributed, or transmitted in any form or by any means, including photocopying; recording, or other electronic or mechanical methods, without the prior written permission of the publisher, except in the case of brief quotations embodied important reviews and certain other noncommercial uses permitted by copyright law. For permission requests, write to the publisher, addressed "Attention: Permissions Coordinator," at the address below.

TABLE OF CONTENT

HOW TO USE THIS BOOK ---7
INTRODUCTION --12
CHAPTER ONE ---16
   UNDERSTANDING SOLAR POWER ---16
   How Does Solar Power Work? --17
   The Evolution of Solar Technology -------------------------------------18
   Key Considerations before Installation --------------------------------19
   Electrical Units: The Language of Energy ------------------------------23
CHAPTER TWO ---30
   BASIC ELECTRICAL PRINCIPLES ---30
   Understanding Voltage, Current, and Resistance ------------------------30
   Understanding AC vs. DC Power ---36
   Advantages and Disadvantages of AC and DC -----------------------------39
   Ohm's Law: Voltage, Current, and Resistance ---------------------------41
   Watt's Law: Understanding Power ---------------------------------------43
   Calculating Power Using Watt's Law ------------------------------------44
   Real-Life Examples and Problem-Solving --------------------------------45
   Understanding Parallel Circuits ---------------------------------------48
   Introduction to Electrical Circuit Elements ---------------------------53
   What is a Resistor? ---53
   Types of Resistors --54
   What is a Capacitor? --55
   What is an Inductor? --57

What is a Transformer? --- 59

Types of Electrical Circuits --- 61

Understanding Kirchhoff's Laws -- 67

Kirchhoff's Voltage Law (KVL) -- 69

Resistive Loads --- 72

Inductive Loads -- 73

Capacitive Loads --- 75

Electrical Shock Hazards -- 84

Importance of Grounding and Bonding -- 86

Introduction to Electrical Measurement Tools ---------------------------------- 90

CHAPTER THREE -- 110

ESSENTIAL TOOLS & EQUIPMENT --- 110

Electrical Safety Guidelines --- 112

Power Tools and Their Applications -- 115

Multi-meter and Clamp Meter -- 122

Solar Racking and Mounting Equipment ------------------------------------ 124

CHAPTER FOUR -- 134

SOLAR PANEL SELECTION & INSTALLATION ----------------------------- 134

Understanding Solar Panel Performance Curves --------------------------- 139

Maximum Power Point Tracking (MPPT) ------------------------------------ 141

Hotspots and Their Prevention --- 145

Optimal Tilt and Azimuth Angles --- 146

Safety Precautions --- 160

Factors Affecting Panel Lifespan --- 166

What to Anticipate from Your Solar Panels over Time ----------------- 169

CHAPTER FIVE --- 172

SOLAR CHARGE CONTROLLERS ------------------------------------- 172

Overview of Charge Controller Roles ---------------------------------- 175

What is MPPT (Maximum Power Point Tracking)? --------------------- 178

When to Use Multiple Controllers -------------------------------------- 193

CHAPTER SIX --- 208

SOLAR BATTERY SYSTEMS --- 208

CHAPTER SEVEN --- 234

OFF-GRID INVERTERS -- 234

CHAPTER EIGHT --- 260

SOLAR WIRING & OVERCURRENT PROTECTION------------------- 260

CHAPTER NINE --- 286

BUILDING YOUR SOLAR SETUP -------------------------------------- 286

Upgrading Components --- 306

CHAPTER TEN -- 308

MONITORING AND MAINTAINING YOUR SOLAR SYSTEM ------------- 308

CHAPTER ELEVEN --- 342

ENERGY STORAGE AND BACKUP SOLUTIONS ----------------------- 342

CHAPTER TWELVE--- 364

FINANCIAL CONSIDERATIONS AND INCENTIVES ------------------- 364

CHAPTER THIRTEEN -- 390

CASE STUDIES AND REAL-WORLD EXAMPLES---------------------- 390

CHAPTER FOURTEEN --- 398

THE FUTURE OF SOLAR TECHNOLOGY--398

HOW TO USE THIS BOOK

Welcome to your in-depth guide to mastering off-grid solar power for beginners. This book is intended to be both an educational resource and a useful tool for rehearsing the complexity of installing and maintaining a solar power system off the grid. Here's a step-by-step guide for making the most of this book.

Understand the Basics

Begin with the Introduction to familiarize yourself with the fundamental principles and problems of off-grid solar power. This part lays the groundwork for what to anticipate from the book and introduces you to the key topics and solutions that will be addressed.

Follow the Chapter Structure

The book is divided into parts that each focus on a different component of off-grid solar power. Here's a quick breakdown on how to go through them.

Solar Power System Basics: Start here if you're new to solar power. We made the foundations of off-grid solar systems self-explanatory be it that you are reading the book as a beginner or as a professional, such as how they function and what factors to consider before installation.

Basic Electricity Rules and formula – we didn't hesitate to review electrical fundamentals, here we explained the key measurements, rules, and formula for understanding how electricity acts in your system.

Tools and Equipment – make you available to know all the necessary tools and equipment before beginning any installation or maintenance work.

Solar Panel Basics: Selection and Installation explore the complication of solar panels, including how to select and install them. Learn about the types of panels, their efficiency, and optimum installation procedures.

We discuss the importance of charge controllers in managing electricity from solar panels to batteries. It describes how to pick and size a charge controller.

Understand the types of batteries used in solar systems, how to maintain them, and how to size them to meet your requirements.

Learn about inverters, which convert DC electricity from solar panels to AC power for your home appliances. Several kinds and their requirements were also reviewed.

Solar Wiring & Overcurrent Protection Devices (OCPD) – Learn how to effectively and safely wire your solar system. We also cover overcurrent protection to avoid electrical problems.

Build a Solar Setup from Scratch – For individuals willing to take on their own solar installation a step-by-step approach on planning and installing a comprehensive solar power system, both theoretical and practical tips were well explained.

Utilize the Case Studies and Examples

Throughout the book, real-life case studies and examples demonstrate typical obstacles and successful solutions. Use these examples to guide your own configuration and troubleshooting.

Apply the Actionable Advice

Each chapter contains useful hints and recommendations. Apply these tips to your solar system to achieve peak performance and efficiency.

Refer to the Resources and Glossary

The book concludes with a glossary of words and further resources for further study. Use them to clarify technical terms and broaden your understanding.

Review the Troubleshooting and Maintenance Tips

Review the troubleshooting and maintenance sections on a frequent basis to ensure that your solar system continues to function properly. These recommendations will assist you in resolving frequent difficulties and keeping your system working properly.

Keep a Notebook

As you read the book, keep a journal to record your observations, questions, and notes on your own solar system. This will help you stay organized and keep track of your progress.

Following this systematic method will provide you with a strong grasp of off-grid solar power and prepare you to construct and operate your own system. Remember, this book is a tool that will provide you with the information and confidence you need to make educated decisions and attain energy independence.

Ready to get started Turn the page and begin your path toward mastering off-grid solar power. Your journey to energy independence starts here.

INTRODUCTION

Imagine waking up to the sound of birds singing, the aroma of fresh morning dew, and the sun rising over a wide, pristine countryside. There is no hum of power lines or distant buzz of a metropolis waking up—only you, nature, and the life you've created. Many of us share the ambition of living a life free of the grid, powered by the sun, and on our own terms.

However, as wonderful as it may seem, the reality of living off-grid brings obstacles that go beyond the idealistic notion of self-reliance. You can find yourself dealing with questions like.

How can I guarantee I have adequate electricity on foggy days? What happens if my batteries fail? How can I match my energy requirements with the capacity of my solar system? These are not merely technical questions; they are the fundamental issues of living off-grid with solar electricity.

In this book, we'll tackle these difficulties straight on. We'll look at the real-world challenges that folks who want to live off the grid encounter, from the earliest planning phases to the day-to-day operation of a solar power system. But, more significantly, we will provide you actual answers, clear counsel, and the courage to face these obstacles.

Let's start with a narrative that may seem familiar. John and Maria, a couple from the city, decided to pursue their longstanding ambition of living off the grid. They invested in a solar power system, delighted by the prospect of independence and sustainability. However, immediately after moving to their secluded cottage, they encountered unanticipated challenges. On overcast days, their power source was unreliable, causing annoyance and uncertainty. Their batteries, which they expected to last years, began to degrade more quickly than anticipated. The excitement of their new life faded as they fought to adjust.

What John and Maria encountered is not unusual. The shift to off-grid life can be fraught with unexpected problems, particularly if you are not completely prepared. But here's the good news: these obstacles can be overcome.

In the following chapters, you'll discover deep insights and practical tips to help you navigate the world of off-grid solar power. We will go over everything from the fundamentals of solar energy to advanced tactics for enhancing your system. You'll learn how to size your solar array, select the correct batteries, and preserve your investment in the long run. We will use case studies and real-life examples to demonstrate potential problems and how to prevent them.

By the conclusion of this book, you'll have the knowledge and skills you need to flourish off the grid rather than just survive. You will be able to take control of your energy demands, lessen your dependency on fossil fuels, and live a sustainable, resilient life driven by the sun.

So, whether you're just starting off or seeking to upgrade your current setup, this book is for you. Let us begin on this adventure together to make the goal of living off the grid a reality.

Are you prepared to take the next step towards energy independence? Let's get started and learn how you may use the power of the sun to build a life of freedom and sustainability. You'll find the answers you've been seeking for, as well as the courage to live the off-grid lifestyle you've always desired.

CHAPTER ONE

UNDERSTANDING SOLAR POWER

Imagine living in a location where the sun provides all of your energy needs—no utility bills, no outages, just pure, sustainable electricity created right in your own backyard. This is the promise of an off-grid solar energy system.

An off-grid solar power system works independently of the standard electrical grid. Off-grid systems, in contrast to grid-tied systems, generate and store all of their own electricity. They are made up of solar panels, a battery bank, a charge controller, and an inverter, all of which work together to turn sunlight into power. This self-sufficient arrangement is perfect for isolated regions, energy independence, or anybody looking to reduce their carbon impact.

However, an off-grid system involves more than just a set of solar panels. It's a gateway to energy independence, allowing you to power your life in accordance with the environment. An off-grid solar power system enables you to live sustainably and freely, whether in a distant cottage, a boat, or an eco-friendly house.

How Does Solar Power Work?

At its core, solar power is about capturing sunlight and converting it into electricity. But how does this process actually work?

It all starts with the solar panels, which are made up of photovoltaic (PV) cells. These cells absorb sunlight and, through a process called the photovoltaic effect, convert it into direct current (DC) electricity. This DC power is then sent to a charge controller, which regulates the voltage and current going into your battery bank, ensuring the batteries are charged efficiently and safely.

Once the electricity is stored in the batteries, it's ready to be used. However, most household appliances run on alternating current (AC) power, not DC. That's where the inverter comes in. The inverter converts the stored DC power into AC power, making it compatible with your home's electrical system.

In essence, solar power allows you to capture the sun's energy, store it, and use it whenever you need it, day or night. It's a clean, renewable energy source that can power your life without the need for fossil fuels or a connection to the grid.

The Evolution of Solar Technology

Solar power technology has come a long way since its early days. Initially developed in the 19th century, solar technology was expensive and inefficient, used mainly for specialized applications like space missions. However, over the past few decades, advances in materials science and manufacturing have dramatically improved the efficiency and affordability of solar panels.

Today, solar power is one of the fastest-growing energy sources worldwide. Innovations in solar panel design, such as the development of thin-film and multi-junction cells, have increased efficiency and reduced costs.

Moreover, the rise of energy storage solutions like lithium-ion batteries has made it possible to store solar power for use when the sun isn't shining, making off-grid systems more viable than ever.

The evolution of solar technology has transformed it from a niche option to a mainstream energy source, empowering more people to go off-grid and take control of their energy needs.

Key Considerations before Installation

Before embarking on an off-grid solar power project, it is good to recognize that this is not a one-size-fits-all approach. Several important variables must be addressed to guarantee that your system satisfies your energy demands and runs efficiently.

1. **Location**: Your geographical location significantly impacts the effectiveness of your solar panels. Areas with high levels of sunlight year-round will generate more power, while regions with frequent

cloud cover or long winters may require a larger system or supplemental energy sources.

2. **System Size**: The size of your system depends on your energy consumption. Installing too small a system may leave you without enough power, while an oversized system could be unnecessarily expensive. Properly sizing your system is crucial for balancing cost and performance.

3. **Battery Storage**: Battery technology is an important component of off-grid systems. The type and size of your battery bank will determine how much energy you can store and for how long. Consider your energy usage patterns and the potential for periods of low sunlight when selecting your batteries.

4. **Costs and Budget**: Off-grid systems can require significant upfront investment. It's important to budget not only for the initial installation but also for ongoing maintenance and potential upgrades.

However, over time, the savings on energy bills and the value of energy independence can offset these costs.

5. **Maintenance**: Unlike grid-tied systems, off-grid solar systems require regular maintenance to ensure they function optimally. This includes checking battery health, cleaning panels, and monitoring system performance. Be prepared for the hands-on responsibility that comes with maintaining your energy source.

Estimating Your Energy Needs

One of the first and most important steps in designing an off-grid solar power system is accurately estimating your energy needs. Understanding how much electricity you use daily will guide you in determining the size of your system.

1. **Assess Your Appliances**: Start by listing all the electrical devices you plan to use, from lights and refrigerators to power tools and electronic gadgets.

For each item, note its wattage and the average number of hours you use it daily.

2. **Calculate Total Energy Consumption**: Multiply the wattage of each device by the hours of use to find its daily energy consumption in watt-hours (Wh). Add up all the watt-hours to get your total daily energy requirement.

3. **Consider Seasonal Variations**: Energy needs can fluctuate with the seasons. For example, heating requirements in winter or increased lighting needs during shorter days can affect your total consumption. Factor in these variations when sizing your system.

4. **Account for Inefficiencies**: No system is 100% efficient. Include a margin of error in your calculations to account for inefficiencies in power conversion, battery storage, and transmission.

5. **Plan for Future Expansion**: If you anticipate your energy needs growing, consider designing

your system with scalability in mind. This could involve installing a larger inverter or leaving space for additional panels and batteries.

Electrical Units: The Language of Energy

Before diving into the technical aspects of off-grid solar power, it's important to understand the basic electrical units that govern how energy is measured and used. These units form the language of electricity, helping you communicate effectively about your system's needs and performance.

- **Watts (W):** The unit of power, representing the rate at which energy is used or generated. It's a product of voltage and current.

- **Ampere (A):** The unit of electric current, which is the flow of electrical charge.

- **Volt (V):** The unit of electric potential or voltage, which drives the current through a circuit.

- **Watt-hour (Wh):** A measure of energy used or generated over time. A watt-hour represents one watt of power utilized for one hour.

Voltage: The Driving Force of Your System

Voltage is the force that pushes electrical current through your system. In an off-grid solar power setup, understanding voltage is critical because it affects every aspect of your system's performance—from how your solar panels generate power to how your batteries store it.

- **High Voltage Systems:** More efficient in transmitting power over longer distances, which reduces energy loss but requires more expensive components.

- **Low Voltage Systems:** Safer and less costly to install but may require thicker wires to handle the current without significant energy loss.

When designing your system, you'll need to choose the appropriate voltage level based on your energy needs and the distance between components.

Current: The Lifeblood of Your Solar System

Current is the flow of electricity, and it's the lifeblood of your solar power system. It's measured in amperes (amps), and it determines how much electricity is being transferred at any given time.

In your system, current is influenced by both the voltage and the resistance in your wiring and components. High current can lead to energy losses, particularly in long cable runs, so it's essential to balance your system's voltage and current to minimize these losses.

Understanding current is crucial for selecting the right wiring, fuses, and other components to ensure your system operates safely and efficiently.

Choosing the Right Mounting Type

Mounting your solar panels is more than just finding a spot and placing them there. The type of mounting you choose can significantly impact the efficiency of your solar system. The goal is to position your panels so that they capture the maximum amount of sunlight throughout the day.

Fixed Mounts: These mounts hold your panels in a stationary position, typically at an angle that is optimized for your location's latitude. Fixed mounts are simple, cost-effective, and require minimal maintenance, but they don't allow you to adjust the panels for seasonal changes in the sun's position.

Adjustable Mounts: These mounts allow you to manually adjust the angle of your panels throughout the year, optimizing their exposure to sunlight. This can increase your system's efficiency, especially in areas with significant seasonal variation in sunlight.

Tracking Mounts: The most advanced option, tracking mounts automatically follow the sun across the sky, ensuring that your panels are always at the optimal angle. While they may greatly boost energy output, they are more costly and require more maintenance.

The right mounting choice depends on your budget, location, and how much you want to maximize your system's efficiency.

Site Selection and Optimization

Selecting the right site for your solar installation is critical. Your panels need to be positioned where they will receive the most sunlight throughout the day, and this often requires careful consideration of shading, orientation, and space.

Shading: Even a small amount of shading can significantly reduce the performance of your solar panels. Look for a site that is free from obstructions like trees, buildings, and other structures that could cast shadows on your panels, especially during peak sunlight hours.

Orientation: In the Northern Hemisphere, solar panels should generally face south to capture the maximum amount of sunlight. However, the exact orientation can be adjusted based on your specific location and energy needs.

Space: Ensure that you have enough space to install the number of panels you need. Consider not only the current size of your system but also potential future expansions.

Next

CHAPTER TWO

BASIC ELECTRICAL PRINCIPLES

Electricity is the lifeblood of every solar power system, and knowing its fundamental principles is essential for harnessing its power properly. In this chapter, we will look at the fundamental concepts of voltage, current, and resistance, as well as how they relate to real-world circumstances. Whether you're new to electrical systems or need a refresher, this chapter will provide you with the information you need to make educated decisions as you set up and manage your off-grid solar power system.

Understanding Voltage, Current, and Resistance

Let's start with the basics. Consider energy to be like water flowing via a pipeline.

- **Voltage** is akin to the water pressure pushing the water through the pipe.

- **Current** is the flow rate of the water moving through the pipe.

- **Resistance** represents any obstruction in the pipe that slows the flow of water.

These three elements are the building blocks of electrical theory, and grasping them is essential for understanding how electricity works in your solar power system.

Voltage (V)

Voltage, measured in volts (V), is the potential difference between two points in an electrical circuit. It is what propels electric current across a circuit. Think of voltage as the force that pushes electrons through the wires of your solar power system. Without sufficient voltage, the current won't flow, and your devices won't function.

Practical Example: Consider a fully charged solar battery. The voltage of this battery might be 12V. This means there is a 12-volt potential difference between the positive and negative terminals of the battery. When you connect a light bulb to this battery, the voltage pushes the electric current through the bulb, causing it to light up.

Current (I)

Current is the movement of electric charge across a conductor and is measured in amperes (A). The pace at which electricity flows through your system is determined by the voltage and resistance in the circuit. The more current flowing through a circuit, the more electricity is being delivered to your devices.

Practical Example: If you have a 12V battery connected to a fan that draws 2 amps of current, the amount of electric charge flowing through the wires to power the fan is 2 amps. The higher the current, the more power your devices receive, but this also means more energy is being drawn from your battery.

Resistance (R)

Resistance, measured in ohms (Ω), is the opposition to the flow of current in an electrical circuit. It's like friction in a pipe—resistance slows down the flow of electricity. Every material has some resistance, but materials like copper, used in wiring, have very low resistance, making them ideal for conducting electricity efficiently.

Practical Example: If you connect a resistor to your 12V battery, the resistor will limit the current flowing through the circuit. This is useful in various applications, such as protecting delicate electronic components from receiving too much current.

Definitions and Units of Measurement

- **Voltage (V):** It refers to the pace at which electricity flows through your system and is determined by the circuit's voltage and resistance. (V).

- **Current (I):** The flow of electrical charge, measured in amperes (A).

- **Resistance (R):** refers to the opposition to current flow, measured in ohms (Ω).

These units are fundamental to electrical calculations and system design. Knowing how to measure and interpret them allows you to troubleshoot issues and optimize your solar power setup.

Relationship between Voltage, Current, and Resistance

The relationship between voltage, current, and resistance is defined by **Ohm's Law**, a key principle in electrical theory. Ohm's Law states:

$$V = I \times R$$

This means that the voltage (V) across a circuit is equal to the current (I) flowing through the circuit multiplied by the resistance (R) in the circuit.

Practical Application: Let's say you have a circuit with a 12V battery and a resistor with a resistance of 6Ω. According to Ohm's Law:

$$I = \frac{V}{R} = \frac{12V}{6\Omega} = 2A$$

So, the current flowing through the circuit would be 2 amps. Understanding this relationship allows you to calculate any one of these values if you know the other two, helping you design circuits that function correctly and safely.

Practical Examples and Applications

Understanding these principles isn't just theoretical; it's essential for practical applications in your off-grid solar power system.

- **Sizing Wires:** The current flowing through your wires must be appropriate for their size to prevent overheating and potential fire hazards. Using Ohm's Law, you can calculate the correct wire size based on the voltage and current in your system.

- **Battery Charging:** When charging batteries with solar panels, it's crucial to match the voltage and current output of the panels to the battery's specifications to ensure efficient charging and avoid damaging the battery.

- **Component Selection:** Choosing the right components, such as resistors, inverters, and charge controllers, depends on your understanding of voltage, current, and resistance. For example, selecting an inverter that can handle the current your devices will draw ensures that your solar system can power all your appliances effectively.

Understanding AC vs. DC Power

When diving into the world of solar power, one of the first concepts you'll encounter is the difference between AC (Alternating Current) and DC (Direct Current) power. This distinction is foundational to understanding how solar power systems work and why specific components are necessary.

AC Power is the type of electricity that flows in alternating directions, meaning it reverses its direction periodically. This type of current is what powers most homes and businesses, delivered via the grid. The standard frequency for AC in most regions is 50 or 60

Hz, meaning it changes direction 50 or 60 times per second.

DC Power, on the other hand, flows in a single, consistent direction. This type of current is what's produced by solar panels and stored in batteries. DC is essential in solar power systems because it allows for the storage of electricity, which can then be converted into AC to power your home.

Differences between AC and DC Power

While both AC and DC power are essential to your off-grid solar system, they serve different functions.

Direction of Flow: AC power alternates direction, while DC power flows in one direction only.

Source: AC is the form of electricity delivered by utility companies. In contrast, DC power is typically generated by batteries, solar panels, and other renewable energy sources.

Transmission: AC power can be transmitted over long distances with minimal loss of energy, making it ideal for grid systems. DC power, however, is more efficient over short distances and is less prone to energy loss during transmission within a system.

Applications of AC and DC in Solar Systems

In an off-grid solar system, both AC and DC power have crucial roles to play:

DC Power Applications: Your solar panels generate DC power. This DC electricity is stored in your battery bank, ensuring you have energy even when the sun isn't shining. DC power is also used directly by some appliances designed for solar power systems, such as LED lighting, certain types of pumps, and DC refrigerators.

AC Power Applications: Most household appliances, from your refrigerator to your television, run on AC power. Therefore, an inverter is necessary to convert the

DC electricity from your batteries into AC power that can be used throughout your home.

Advantages and Disadvantages of AC and DC

Advantages of DC Power

1. **Efficiency:** DC power is efficient for storage and use in smaller systems. Solar panels produce DC, which can be directly stored in batteries without conversion losses.

2. **Simplicity:** DC systems are generally simpler and less complex, making them easier to set up for off-grid applications.

3. **Compatibility:** DC power is inherently compatible with renewable energy systems like solar, making it a natural fit for off-grid living.

Disadvantages of DC Power:

1. **Limited Range:** DC power loses efficiency over long distances, making it less suitable for grid applications.

2. **Inverter Requirement:** To use DC power for AC appliances, you must convert it, which can result in some energy loss and adds complexity to your system.

Advantages of AC Power:

1. **Long-Distance Transmission:** AC power is ideal for transmitting electricity over long distances with minimal loss, which is why it's used by the grid.

2. **Widespread Use:** Most appliances and devices are designed to run on AC power, making it the standard for household use.

Disadvantages of AC Power:

1. **Complexity:** Converting solar-generated DC power to AC involves additional components, such

as inverters, which can increase the complexity and cost of your system.

2. **Conversion Loss:** The process of converting DC to AC involves energy loss, which can reduce overall system efficiency.

Ohm's Law: Voltage, Current, and Resistance

Ohm's Law is a key principle in the field of electricity. It explains the interaction of three important elements: voltage (V), current (I), and resistance (R).

Ohm's Law says that the current flowing through a conductor between two locations is exactly proportional to the voltage across the two sites and inversely proportional to their resistance. This relationship is expressed through the formula:

$$V = I \times R$$

Voltage (V): The force that pushes electric charge through a conductor. It's measured in volts (V).

Current (I): Current (I): The flow of electrical charge, measured in amperes (A).

Resistance (R): is the opposition to the passage of current, measured in ohms (Ω).

Detailed Explanation of Ohm's Law

Imagine you're pushing a sled across the snow. The voltage is like the force you apply to the sled—the more force you apply, the faster it moves (higher current). However, if the snow is deep and heavy (higher resistance) you'll have to push much harder to move the sled the same distance. This is the essence of Ohm's Law.

In practical terms, if you increase the voltage in a circuit while keeping the resistance constant, the current will increase. Conversely, if you increase the resistance while keeping the voltage the same, the current will decrease. Understanding this relationship is crucial when designing a solar power system, as it affects how efficiently your system operates.

For example, if you're running wires from your solar panels to your battery bank, the resistance in the wires (due to length, thickness, and material) will impact how much current flow through them. If the resistance is too high, you might lose power as heat, reducing the efficiency of your system.

Watt's Law: Understanding Power

While Ohm's Law helps you understand the relationship between voltage, current, and resistance, Watt's Law helps you calculate power, which is the rate at which energy is used or produced. Power is measured in watts (W), and Watt's Law is expressed as:

$$P = V \times I$$

- **Power (P):** The rate of energy consumption or production, measured in watts (W).

- **Voltage (V):** As before, this is the force driving the current, measured in volts (V).

- **Current (I):** Is the flow of electrical charge, it's measured in amperes (A).

Calculating Power Using Watt's Law

Watt's Law is very important for determining how much electricity your solar panels generate and how much power your appliances require.

For example, if your solar panel generates 20 volts and 5 amps of current, the power output is:

$$P = 20V \times 5A = 100W$$

This signifies that your solar panel generates 100 watts of power under these conditions. In contrast, if you have a light bulb that consumes 60 watts of power and runs at 120 volts, you may compute the current it draws:

$$I = \frac{P}{V} = \frac{60W}{120V} = 0.5A$$

This indicates that the lightbulb requires 0.5 amps of electricity from the circuit.

Real-Life Examples and Problem-Solving

Let's apply these laws to real-life scenarios you might encounter when setting up your off-grid solar power system.

Scenario 1: Choosing the Right Wire Size

You're installing a solar power system and need to run cables from your solar panels to the charge controller. The cables have resistance, and if they are too thin or too lengthy, they may lose power through heat. Using Ohm's Law, you may calculate the voltage drop between the wires and decide if bigger cables are necessary to minimize resistance and assure maximum power transfer.

Scenario 2: Sizing Your Solar System

You intend to power a cabin with a few lights, a refrigerator, and a fan. First, apply Watt's Law to determine overall power usage. Let's assume the lights use 100 watts, the refrigerator 200 watts, and the fan 50 watts. To calculate the total power required, multiply 100W, 200W, and 50W to get 350W. Next, establish how many solar panels you'll need by calculating each panel's power output and ensuring it meets or surpasses the overall power usage.

Series and Parallel Circuit Configurations

Understanding the distinction between series and parallel configurations is critical when working with electrical circuits in a solar power system. These two main types of circuits control how voltage and current are distributed between your components, affecting your system's overall performance and efficiency.

Understanding Series Circuits

In a series circuit, components are linked end-to-end in a single route to allow electric current to flow. Consider a strand of traditional Christmas lights: if one bulb burns out, the entire string stops operating. This is because the current has only one path to go, and a break anywhere in the circuit interrupts the flow of energy.

Key Characteristics of Series Circuits:

- **Voltage:** The overall voltage across a series circuit is the sum of the voltages across each component. This implies that if you add more components (such as resistors or batteries), the overall voltage rises.

- **Current:** The current in a series circuit is the same through all components. Because there is only one path for the current to take, every component experiences the same amount of current.

- **Resistance:** In a series circuit, the overall resistance is calculated by adding the individual resistances together. When too many components are added, the voltage might drop significantly.

Example:
If you connect two 12V batteries in series, the total voltage delivered to your circuit is 24V (12V plus 12V). However, if you add more resistance-containing

components, the overall current may drop as the total resistance increases.

Understanding Parallel Circuits

A parallel circuit has components linked across the same two locations, resulting in many routes for electric current to flow. Unlike in a series circuit, if one component fails in a parallel circuit, the others continue to function since each has its own independent route.

Key Characteristics of Parallel Circuits:

- **Voltage:** In a parallel circuit, the voltage across each component is the same as the voltage at the source. This means that each component receives the entire voltage, regardless of the other.

- **Current:** In a parallel circuit, the total current is the sum of the currents flowing through each branch. The current might vary across branches based on the resistance of each route.

- **Resistance:** The overall resistance in a parallel circuit is smaller than the lowest individual resistance in any branch. Adding additional branches lowers the total resistance, enabling more current to flow.

Example:

When you connect two 12V batteries in parallel, the voltage stays constant, but the available current capacity doubles. This configuration is frequently used in solar systems to increase the available current without affecting the voltage, allowing more power-hungry devices to operate concurrently.

Combining Series and Parallel Circuits

To maximize performance, real-world solar power systems sometimes include both series and parallel circuits. For example, you might link solar panels in series to increase the total voltage (which is beneficial for reducing energy loss over long distances), and then join those series strings in parallel to increase the total current (which provides more power to your devices).

Series-Parallel Hybrid Configuration

- **Higher Voltage, Higher Current:** Combining series and parallel setups allows for increased voltage and current, improving power output and ensuring your system fulfills the energy demands of your off-grid arrangement.

- **Improved Reliability:** A series-parallel design enhances your system's dependability. If one component fails in a parallel branch, the others will continue to work, ensuring overall system performance.

Example:

In a solar array, three strings of four panels each may be linked in series, followed by parallel connections. This arrangement allows you to raise both voltage and current, creating a more balanced and efficient power output.

Practical Examples of Series and Parallel Configurations

To bring these principles to life, let's look at some practical instances.

1. **Solar Panel Array Configuration:**

    Series Configuration: Suppose you have four 100W solar panels, each with an output of 12V and 8.3A. By connecting these panels in series, you would get a total of 48V (12V x 4) while maintaining the current at 8.3A. This is useful when you need higher voltage to reduce losses over long distances or to match the voltage requirements of certain inverters.

    Parallel Configuration: If you connect the same four 100W panels in parallel, the voltage would remain at 12V, but the current would increase to 33.2A (8.3A x 4). This setup is ideal for charging a battery bank that requires higher current for faster charging.

2. **Battery Bank Configuration:**

   Series Configuration: Connecting two 6V batteries in series gives you 12V, suitable for standard 12V inverters.

   Parallel Configuration: Connecting two 6V batteries in parallel keeps the voltage at 6V but doubles the amp-hour capacity, which extends the usage time before recharging is needed.

3. **Lighting Circuit:**

   Series Configuration: In a simple series circuit with multiple light bulbs, if one bulb burns out, the circuit is broken, and all lights go out.

   Parallel Configuration: In a parallel circuit, each light bulb operates independently. If one burns out, the others continue to shine brightly.

Introduction to Electrical Circuit Elements

Understanding the fundamentals of electricity is essential for operating any off-grid solar power system. Circuit elements are the vital components that make up every electrical system. Resistors, capacitors, inductors, and transformers are the components that control, store, and transport electrical energy in your system. In this chapter, we'll go over each of these components, discussing their purposes, kinds, and real-world applications, so you can make educated decisions about creating and maintaining your off-grid system.

Resistors: Function and Types

What is a Resistor?

A resistor is one of the simplest yet most important components in any electrical circuit. Its principal role is to limit or regulate the flow of electric current, ensuring that the other components in the circuit work safely. Without resistors, circuits are quickly overloaded, resulting in system damage or failure.

How Do Resistors Work?

When an electrical current flows through a resistor, it inhibits the flow of electrons, causing energy to be lost as heat. Resistance is measured in ohms (Ω) and directly influences the amount of current permitted to travel through the circuit.

Types of Resistors

1. **Fixed Resistors**: These have a fixed resistance value and are utilized in circuits where constant current control is required.

2. **Variable Resistors (Potentiometers)**: These have adjustable resistance, allowing you to alter the current flow in a circuit. They are widely utilized in devices such as volume controls and light dimmers.

3. **Thermistors**: A resistor whose resistance varies with temperature. They are commonly utilized in temperature sensing and protection circuits.

4. **Light-Dependent Resistors (LDRs)**: These resistors alter resistance in response to light intensity. They are valuable in light-sensitive applications, such as automated street lighting.

Capacitors: Function and Types

What is a Capacitor?

Capacitors are valued components in electrical circuits, typically used to store and discharge electrical energy. They briefly store electrical charge and then discharge it when needed, making them vital in smoothing out oscillations in power supply and producing energy bursts when needed.

How Do Capacitors Work?

A capacitor is made up of two conducting plates separated by an insulating substance known as the dielectric, when a voltage is put between the plates, an electric field forms, and the capacitor stores energy in this field. Capacitance, measured in farads (F), determines the amount of energy a capacitor can hold.

Types of Capacitors

1. **Ceramic Capacitors**: Known for their compact size and reliability, these capacitors are often employed in high-frequency circuits.

2. **Electrolytic Capacitors**: These have a larger capacitance value and are used in power supply circuits to smooth out voltage swings. However, they are polarized therefore they must be linked in the right way.

3. **Tantalum Capacitors**: Similar to electrolytic capacitors, but with superior performance in terms of size-to-capacitance. They're also polarized and employed in delicate situations with little area.

4. **Film Capacitors**: These provide excellent accuracy and stability, making them perfect for applications that need minimal noise and good dependability, such as audio equipment.

Inductors Function and Types

What is an Inductor?

An inductor is a passive component that stores energy in a magnetic field when an electrical current passes through it. Inductors are important for filtering and controlling the flow of AC (alternating current) in circuits, helping to block or pass particular frequencies as needed.

How Do Inductors Work?

Inductors are generally constructed up of a wire coil that generates a magnetic field when current flows through it. The inductor resists changes in the current passing through it by producing a voltage that opposes these changes, a property known as inductance, measured in henries (H).

Types of Inductors

1. **Air-Core Inductors**: These are basic wire coils without a magnetic core that are utilized in high-frequency applications requiring minimal inductance.

2. **Iron-Core Inductors**: Inductors with an iron core have a larger inductance and are utilized in low-frequency applications such as transformers and power supply.

3. **Toroidal Inductors**: These doughnut-shaped inductors, which are very efficient, are employed in circuits where space is restricted and electromagnetic interference needs to be reduced.

4. **Variable Inductors**: These enable for the change of inductance by sliding the core in and out of the coil, giving flexibility in tuning circuits.

What is a Transformer?

Transformers are important components in electrical systems, particularly in off-grid solar systems, where they are used to step up (raise) or step down (lower) voltage levels to meet the demands of various system components. They operate on the basis of electromagnetic induction and are critical for properly distributing power across several devices.

How Do Transformers Work?

A transformer is made up of two wire coils, called primary and secondary windings that are around a magnetic core. When AC voltage is delivered to the main winding, it generates a magnetic field that induces voltage in the secondary winding. The number of turns of wire in each coil determines the voltage ratio of the main and secondary windings.

Applications of Transformers

1. **Power Distribution**: Transformers are frequently employed in electricity networks to step up voltage for long-distance transmission and step down voltage for safe usage in homes and businesses.

2. **Isolation**: They can separate distinct areas of a circuit, shielding delicate components from voltage spikes and interference.

3. **Voltage Matching**: Transformers help to match the voltage levels of various devices in a system, ensuring that each component receives the proper voltage.

4. **Impedance Matching**: Transformers are employed in audio and radio-frequency applications to match the impedances of different circuit parts, optimizing power transmission while reducing signal loss.

Types of Electrical Circuits

An electrical circuit is a channel through which electricity may travel. There are three sorts of circuits that you must understand: open circuits, closed circuits, and short circuits. Each behaves differently and has its own impact on your solar power system.

Open Circuits: Causes and Effects

An open circuit arises when the channel for electricity is incomplete and current cannot flow. Consider it as a broken bridge that automobiles (electrons) cannot traverse. This might occur in a solar power system if there is a loose wire, a disconnected component, or a switch turned off.

Causes:

- **Loose or disconnected wires:** A common cause of open circuits in off-grid systems.
- **Faulty switches or connectors:** If a switch or connector fails, it can break the circuit.

- **Corrosion:** Over time, connections can corrode, leading to an open circuit.

Effects:

- **No power flow:** The most immediate effect of an open circuit is that your solar system won't work as expected. Devices won't receive power.

- **System inefficiencies:** Even if only part of the system is affected, it can lead to inefficiencies, reducing the overall performance.

Recognizing the signs of an open circuit and addressing them promptly is crucial to maintaining a reliable power supply.

Closed Circuits: Definition and Functionality

A **closed circuit** is a complete electrical loop that allows current to flow uninterrupted from the power source to the load and back again. In a well-functioning solar power system, circuits are designed to be closed so that power flows smoothly to where it's needed.

Functionality:

- **Power distribution:** Closed circuits ensure that electricity reaches every component of your system, powering your lights, appliances, and other devices.

- **Controlled operation:** With properly closed circuits, you can control the flow of electricity using switches, timers, or automated systems.

Reliability: A properly closed circuit is the hallmark of a reliable electrical system. Ensuring that your circuits are complete and functioning well is key to avoiding power interruptions.

Short Circuits: Causes, Effects, and Prevention

A **short circuit** occurs when electricity takes an unintended path, often due to a fault in the wiring. This can cause a surge of electricity that bypasses the intended load, leading to overheating, damage, or even fire. Short circuits are one of the most dangerous electrical problems.

Causes:

- **Damaged insulation:** When wires lose their protective insulation, they can come into contact with each other or other conductive materials.

- **Faulty wiring:** Improperly installed wiring can lead to short circuits, especially if wires are pinched or cut.

- **Water exposure:** Water can create a conductive path between wires, leading to a short circuit.

Effects:

- **Overheating:** Short circuits can cause components to overheat, which can lead to fires.

- **Damage to components:** The sudden surge of current can damage solar panels, inverters, or batteries.

- **System failure:** A short circuit can cause a total system shutdown, leaving you without power until the issue is resolved.

Prevention:

- **Regular inspections:** Regularly check your wiring for signs of wear or damage.

- **Use quality materials:** Ensure all wiring and components are of high quality and properly rated for your system.

- **Install circuit protection devices:** Fuses, breakers, and ground fault protection devices can help prevent the severe consequences of a short circuit.

Circuit Protection: Fuses, Breakers, and Ground Fault Protection

Protecting your electrical circuits is essential to prevent damage and ensure safety. Here's how you can safeguard your off-grid solar power system:

Fuses:

- **How they work:** Fuses are simple devices that break the circuit when too much current flows through, preventing overheating and potential fires.

- **Where to use them:** Fuses are often used in smaller circuits where the risk of a current surge is high.

Breakers:

- **How they work:** Circuit breakers serve a similar function to fuses but can be reset after they trip. They break the circuit when they detect a surge of current, protecting your system from damage.

- **Where to use them:** Breakers are commonly used in main distribution panels and for protecting larger circuits.

Ground Fault Protection:

- **How it works:** Ground fault protection devices detect when electricity is flowing along an unintended path, such as through water or a person. They cut off power to prevent electric shocks.

- **Where to use them:** These devices are essential in areas where water is present, such as near solar panel connections or battery storage areas.

Understanding Kirchhoff's Laws

Solid grasp of basic electrical principles is important. Among these principles, Kirchhoff's Laws stand out as foundational concepts that govern how current and voltage behave in electrical circuits. Named after the German physicist Gustav Kirchhoff, these laws are the bedrock upon which much of electrical engineering is built.

Kirchhoff's Current Law (KCL) states that the total current entering and leaving a junction (or node) in an electrical circuit are equal. In other words, the amount of currents entering a node must match the sum of currents exiting. This is based on the charge conservation principle, which states that charge (and hence current) cannot be generated or destroyed at any node.

Mathematically:

$$\sum I_{\text{in}} = \sum I_{\text{out}}$$

Where I_{in} is the current flowing into the node and I_{out} is the current flowing out.

Example: Consider a simple circuit with three wires connecting at a junction. If 5 amps flow into the junction through one wire and 2 amps flow in through another, KCL calculates that a total of 7 amps must flow out through the third wire. This rule is important for examining circuits, especially in understanding how current is dispersed.

Kirchhoff's Voltage Law (KVL)

Kirchhoff's Voltage Law (KVL) asserts that the total of all electrical potential differences (voltages) surrounding any closed loop in a circuit equals zero. This reflects the conservation of energy principle, which states that the energy given to a charge while going around a loop must equal the energy used up in overcoming resistances in that loop.

Mathematically

$$\sum V_{\text{drops}} = \sum V_{\text{gains}}$$

Simply said, the total voltage obtained by charges as they go through a loop (by sources such as batteries) must match the total voltage decreased across the loop's resistive parts.

Example: Consider a circuit loop that includes a 12-volt battery and three resistors in series. If the voltage dips across these resistors total 12 volts, the battery's power is fully used within the loop, and the loop meets KVL standards. If your measures don't add up, there's something wrong with the setup or the computations.

Applications of Kirchhoff's Laws in Circuit Analysis

Kirchhoff's Laws are useful tools in studying electrical circuits, especially in complicated systems where merely using Ohm's Law is insufficient.

1. Solving Complex Circuits: Off-grid solar systems include several interrelated components, including solar panels, batteries, inverters, and loads. Using KCL and KVL, you can solve for unknown currents and voltages in these complicated networks, ensuring that each component functions within safe bounds.

2. Ensuring Efficient Energy Distribution: Kirchhoff's Laws help you understand how energy flows across a circuit, which is important for increasing the efficiency of any solar power system. By evaluating current and voltage in various portions of the system, you can detect and repair inefficiencies, ensuring that as much solar energy as possible is captured and sent to your appliances.

3. Troubleshooting System Issues: When your solar system isn't performing properly, Kirchhoff's Laws can help you diagnose the issue. Whether it's a defective connection or an overloaded component, these principles assist track the flow of current and voltage distribution, allowing you to determine where things went wrong.

Real-World Example: Imagine you've installed a new off-grid solar system. After installation, you discover that your batteries aren't charging properly. Using KCL, you may determine whether the current entering the battery matches the predicted output from the charge controller. KVL may also be used to determine whether the voltage across the batteries matches the output voltage of your solar panels. If disparities occur, these regulations will help you discover and solve the issue—whether it's a bad connection, a defective charge controller, or a wiring problem.

Resistive Loads

Resistive loads are perhaps the easiest to comprehend and most prevalent sort of electrical load. When an electric current travels through a resistive load, it is completely transformed into heat and, occasionally, light.

Examples of resistive loads include:

- **Incandescent light bulbs**: These bulbs generate light by heating a tungsten filament until it glows.

- **Electric heaters**: These convert electrical energy directly into heat.

- **Toasters**: Similarly, a toaster uses electrical resistance to produce heat that toasts bread.

Characteristics of resistive loads

- **Current and voltage in phase**: In a resistive load, the current and voltage are perfectly in phase, meaning they reach their maximum and minimum values simultaneously.

- **No power factor correction required**: Resistive loads do not cause any reactive power issues, so there's no need for power factor correction.

- **Consistent behavior in AC and DC**: Whether in an AC or DC circuit, resistive loads behave consistently, making them predictable and easy to manage.

Inductive Loads

Inductive loads are more complex, as they involve devices that create a magnetic field when electricity flows through them. This type of load stores energy in the form of a magnetic field, which can cause a delay between the current and voltage.

Examples of inductive loads include:

- **Electric motors**: Found in appliances like refrigerators, fans, and pumps, these motors rely on magnetic fields to create motion.

- **Transformers**: Used to change voltage levels in electrical circuits, transformers operate on the principle of induction.

- **Fluorescent lighting**: These lights use inductive ballasts to regulate current.

Characteristics of inductive loads

- **Current lags behind voltage**: In inductive loads, the current lags behind the voltage due to the time it takes for the magnetic field to build up and collapse.

- **Reactive power**: Inductive loads consume reactive power in addition to real power, which can cause inefficiencies in power systems.

- **Power factor considerations**: Inductive loads typically have a low power factor, meaning they use more current than resistive loads for the same amount of power. This can lead to higher losses in the system and may require power factor correction equipment.

Managing inductive loads in an off-grid solar system requires careful consideration of their starting currents and their effect on the overall efficiency of the system.

Capacitive Loads

Capacitive loads are less common but still important in certain applications. These loads store electrical energy in an electric field, usually involving components like capacitors.

Examples of capacitive loads include:

- **Capacitor banks**: Used in industrial settings for power factor correction.
- **Power factor correction devices**: These are designed to counteract the effects of inductive loads.
- **Certain types of LED drivers**: These can exhibit capacitive characteristics.

Characteristics of capacitive loads

- **Current leads voltage**: In capacitive loads, the current leads the voltage, which is the opposite of what happens in inductive loads.

- **Can improve power factor**: Capacitive loads are often used to improve the power factor in a system with inductive loads.

- **Potential resonance issues**: If not properly managed, capacitive loads can cause resonance in AC circuits, leading to voltage spikes and system instability.

In off-grid systems, capacitive loads are typically used to balance inductive loads and improve overall system performance.

Load Behavior in AC and DC Circuits

The behavior of loads can vary significantly depending on whether they are connected to an AC or DC circuit:

- **AC Circuits**: Alternating current changes direction periodically, which affects how inductive and capacitive loads behave. In AC circuits, inductive and capacitive loads can cause power factor issues that need to be managed carefully. Understanding the phase relationships between current and voltage in these circuits is essential for designing efficient systems.

- **DC Circuits**: Direct current flows in one direction, which simplifies the behavior of loads. In DC circuits, resistive loads behave in a straightforward manner, while inductive loads can cause issues during switching operations due to the stored magnetic energy. Capacitive loads in DC circuits are generally used for filtering and stabilizing voltage.

Knowing the difference in load behavior between AC and DC circuits helps in designing the appropriate system components and wiring for your off-grid solar setup.

Power Factor and Its Importance

Among these principles, Power Factor stands out as a key element that can significantly impact the performance and cost-effectiveness of your solar setup. We'll break down what power factor is, why it matters, and how you can manage it to get the most out of your solar system.

Understanding Power Factor

Consider your solar power system to be a rowing crew on a boat. To propel the boat ahead efficiently, everyone must row in perfect harmony. However, if any rowers are out of rhythm, some of their effort is lost in misalignment rather than contributing to the boat's forward movement. This analogy helps to clarify the notion of power factor.

Power Factor (PF) is a measure of how efficiently electrical power is used in your system. It is the ratio of usable power (measured in watts) to perceive power (measured in volt-amperes) delivered to the circuit. Simply said, power factor measures how successfully your electrical equipment turns electricity into productive labor.

- **Real Power (P):** The energy that really drives your equipment and does valuable tasks.

- **Apparent Power (S):** The entire power generated by your solar system, including both actual and reactive power.

- **Reactive Power (Q):** The electricity goes back and forth between the source and the load, performing no productive work.

A power factor of one, or unity, indicates that all available power is being used effectively. However, in real-world circumstances, power factor often varies from 0 to 1. A lower power factor implies inefficiency in your system, which means you're taking more power than is required for your gadgets to operate effectively.

Power Factor Correction

Why should you worry about the power factor? A poor power factor might have various negative effects on your off-grid solar system. They include:

- **Increased Energy Losses:** More current is needed to produce the same amount of useable power, resulting in increased losses in electrical lines and components.

- **Overloaded Components:** Electrical components like inverters, generators, and batteries may be required to handle more power than they are rated for, which can lead to overheating and premature failure.

- **Reduced System Efficiency:** Your solar panels and batteries may be producing and storing energy that is being inefficiently used, lowering the overall efficacy of your system.

Power Factor Correction (PFC) is the process of increasing the power factor of your solar system. This may be accomplished by adding capacitors or inductors to the electrical circuit. These devices serve to mitigate the impacts of reactive power, increasing the total power factor.

There are two primary forms of power factor correction.

1. **Static Power Factor Correction:** This entails connecting capacitors in parallel with the equipment that produces a low power factor. It is primarily used for single devices or tiny systems.

2. **Automatic Power Factor Correction (APFC):** This is a more complex solution in which a controller automatically changes the power factor by turning capacitors on and off as necessary. This strategy is best suited for bigger systems with variable loads.

Impact of Power Factor on Solar Systems

Power factor is important in off-grid solar systems since it affects overall efficiency and lifetime. Here's how.

Battery Life: A low power factor might increase demand on your batteries, shortening their lifespan. Improving your power factor ensures that your batteries only give the necessary power, extending their life.

Inverter Efficiency: Inverters, which convert DC electricity from solar panels to AC power for appliances, are sensitive to power factor. A low power factor can require inverters to work harder, lowering efficiency and increasing the risk of failure.

System Sizing: When constructing an off-grid solar system, a low power factor indicates that larger components (such as solar panels, inverters, and batteries) would be required to satisfy power requirements. This may raise the initial cost of your system. Correcting the power factor enables a more precisely sized system, maximizing both cost and performance.

Energy Savings: Ultimately, adjusting the power factor reduces the quantity of energy lost in your system. This not only reduces your power expenses, but also guarantees that your solar system is as green and efficient as possible.

Basic Electrical Safety Principles

Electricity, while a strong and necessary energy source, can be extremely deadly if not handled correctly. The first and most important step in any electrical project is to emphasize safety. Here are the basic safety guidelines that you should follow:

1. **Turn off Power before Working:** Always turn off electricity before doing any repairs on your solar system. This is the most basic, yet vital, guideline for avoiding electrical shocks or inadvertent short circuits.

2. **Use Proper Tools and Equipment:** Make sure you're using insulated tools and wearing proper safety gear, such gloves and goggles. This reduces the likelihood of inadvertent contact with live wires.

3. **Avoid Water:** Never operate with electrical components when damp or with moist hands. Water conducts electricity and can cause significant shocks or damage to your equipment.

4. **Follow Manufacturer Instructions:** Always follow the manufacturer's instructions for installation, operation, and maintenance of your solar equipment. These instructions are designed to guarantee maximum performance and safety.

5. **Regular Inspections:** Inspect your system on a regular basis for wear and damage, as well as any threats. Catching faults early can avoid accidents and extend the life of your system.

Electrical Shock Hazards

Electrical shock occurs when your body becomes a part of an electrical circuit, allowing electricity to flow through you. The severity of an electric shock can range from a minor tingle to a fatal injury, depending on the current's path, duration, and intensity. Here's what you need to know about shock hazards.

1. **Types of Shocks:**

    Direct Shock: This happens when you directly touch a live wire or component, allowing current to pass through your body.

    Indirect Shock: Occurs when you touch a conductive object that is not supposed to be live, but has become electrified due to a fault in the system.

2. **Symptoms of Electric Shock:** Depending on the severity, symptoms can include tingling, muscle contractions, burns, or even cardiac arrest. Immediate medical attention is crucial in cases of severe shock.

3. **Prevention:** The key to preventing electrical shocks is to always assume that wires are live until proven otherwise, use appropriate protective gear,

and double-check that power is off before starting any work.

Importance of Grounding and Bonding

Grounding and bonding are critical concepts in electrical systems, especially for solar power setups. Proper grounding ensures that in the event of a fault, excess electricity has a safe path to the ground, reducing the risk of shock, fire, or damage to your equipment.

1. **Grounding Basics:**

    Purpose: Grounding provides a safe pathway for electricity to dissipate into the earth, preventing unwanted build-up that could lead to dangerous conditions.

    Methods: Grounding can be achieved by connecting the system's metal parts to the earth using grounding rods or plates. This connection ensures that any stray electricity doesn't accumulate on surfaces that could be touched.

2. **Bonding Basics:**

   Purpose: Bonding ensures that all metal parts that could potentially carry electricity are connected together and to the ground. This prevents dangerous voltage differences between these parts.

   Importance: Bonding is essential for both safety and system performance, as it ensures that in the event of a fault, the electricity is safely directed to the ground.

3. **Common Grounding and Bonding Issues:**

   Loose Connections: Over time, connections can become loose, leading to ineffective grounding. Regular checks are necessary.

   Corrosion: In outdoor systems, corrosion can compromise the integrity of grounding and bonding connections. Use corrosion-resistant materials and inspect regularly.

Use of Insulation and Protective Devices

Insulation and protective devices are your first line of defense against electrical hazards. These components ensure that electricity flow only where it's supposed to, preventing accidental contact and reducing the risk of short circuits or fires.

1. **Electrical Insulation:**

    Function: Insulation is the material that covers wires, preventing the current from escaping or coming into contact with other conductive materials. This keeps the electricity contained within its designated path.

    Types of Insulation: Common materials include plastic, rubber, and Teflon. The type used depends on the voltage and environment in which the wiring is installed.

2. **Protective Devices:**

   Circuit Breakers: These devices automatically cut off the flow of electricity if they detect an overload or short circuit, preventing potential damage or fire.

   Fuses: Fuses serve a similar purpose by breaking the circuit when the current exceeds a safe level. Unlike breakers, fuses need to be replaced once they've blown.

   Ground Fault Circuit Interrupters (GFCIs): These are critical in preventing shocks. GFCIs monitor the current flowing through a circuit and shut it off immediately if they detect any imbalance, which could indicate a potential shock hazard.

3. **Installation and Maintenance:**

Proper Installation: Ensure that all insulation and safety devices are appropriately placed according to the manufacturer's instructions and the local electrical code.

Regular Testing: Periodically test circuit breakers, GFCIs, and other protective devices to verify they are working properly.

Introduction to Electrical Measurement Tools

Whether you're troubleshooting, installing, or optimizing your solar system, having the correct tools and understanding how to use them can be the difference between success and aggravation.

Electrical measuring instruments are important for determining the performance and safety of your system. They let you to measure voltage, current, resistance, and more, ensuring that your solar power system performs effectively and safely.

We will introduce you to the most often used tools in the field, explaining their purposes, how to use them efficiently, and when they are most useful.

Usage and Applications

A multimeter is perhaps the most versatile instrument in any electrician's toolbox. It integrates numerous measuring capabilities in one unit—usually voltage (both AC and DC), current, and resistance.

How to Use a Multimeter

1. **Measuring Voltage:** Set the multimeter's voltage range (AC or DC). Place the probes at the places you want to measure. The voltage measured between the two spots will be shown on the screen.

2. **Measuring Current:** To measure current, the circuit must be interrupted so that the multimeter may be connected in series to the load. To avoid damage, make sure the multimeter's current range is adjusted correctly.

3. **Measuring Resistance:** Set the multimeter to resistance mode (ohms). Touch the probes together to confirm the multimeter is operational (it should

read zero or near to it). Next, set the probes on the component or cable that you wish to measure.

Applications

- **System Troubleshooting:** Use a multimeter to identify problems in your solar system by determining whether the voltage is within normal limits or whether a wire is connected.
- **Battery Health:** Check your batteries' voltage on a regular basis to verify that they are charging and draining properly.
- **Panel Output:** Measure the voltage and current straight from your solar panels to ensure that they are providing power as planned.

Oscilloscopes: Basics and Applications

An oscilloscope may appear to be a sophisticated instrument, yet it is extremely useful for seeing the waveforms of electrical signals, especially in systems with complicated or fluctuating currents and voltages.

Basics of an Oscilloscope

- **What It Does:** An oscilloscope shows a graph of voltage over time, allowing you to examine how it varies. This is especially helpful for evaluating AC signals, pulsating DC signals, and other time-varying electrical values.

- **How to Use It:** Connect the oscilloscope probes to the circuit locations you want to measure. The screen will display the voltage waveform. Adjust the time base and voltage scale to gain a better perspective of the signal.

Applications

- **Waveform Analysis:** Examine the waveform to determine inverter output or PWM (Pulse Width Modulation) controller performance.

- **Signal Troubleshooting:** Identify problems with noisy signals or unexpected variations in your system.

- **Component Testing:** Use an oscilloscope to examine the performance of capacitors, transistors, and other components by studying their response to input signals.

Clamp Meters Functionality and Use

Clamp meters are a rapid and non-invasive technique to measure current, making them helpful for diagnosing faults in active circuits without having to unhook anything.

How Clamp Meters Work

- **Non-Contact Measurement** Clamp meters measure current by clamping onto a conductor. They detect the magnetic field caused by the current passing through the wire and convert it to a readable current value.

- **Safety** Clamp meters are a safer technique to measure current since they do not need direct contact with the wire. This is especially true for high-power applications.

Applications

- **Current Measurement:** Quickly monitor the current flowing through cables in your solar system, whether they're linked to panels, batteries, or loads.

- **Load Balancing:** Measure the current in each branch of the circuit to ensure that your system is appropriately balanced.

- **Fault Detection:** Detect overloaded circuits or unusual current draws that may indicate problems or inefficiency.

Circuit Diagrams and Symbols

Understanding circuit diagrams and the symbols used in them is essential for developing and debugging electrical systems. A circuit diagram is a visual depiction of an electrical circuit. It shows how the components are linked together.

Key Symbols

> **Resistor:** The zigzag line indicates where resistance is placed in the circuit.
>
> **Capacitor:** Two parallel lines representing a capacitor, which is a temporary storage device for electrical energy.
>
> **Inductor:** A series of loops or a coil symbolizes an inductor, indicating where magnetic fields are created to oppose changes in current.
>
> **Battery:** A pair of lines, one longer than the other, represents a battery, with the longer line indicating the positive terminal.
>
> **Switch:** is a device that can open and close a circuit, hence regulating the flow of electricity.

Ground: A set of three lines decreasing in size, resembling a triangle, indicates the ground, the point of reference for the circuit's voltage.

Reading Circuit Diagrams

Follow the Flow: Start from the power source and follow the flow of electricity through the circuit. Understanding the direction of current flow helps in diagnosing issues.

Identify Components: Use the symbols to identify each component and understand its role within the circuit.

Check Connections: Ensure that all connections are correct according to the diagram before powering up your system. Misconnections can lead to shorts, damage, or inefficiencies.

Applications

System Design: Use circuit diagrams to plan out your solar setup, ensuring all components are properly connected and compatible.

Troubleshooting: When something goes wrong, refer to the circuit diagram to trace the issue and identify faulty components or connections.

Learning and Communication: Circuit diagrams are a universal language in electronics. Understanding them allows you to communicate more effectively with others in the field and to learn from existing designs.

Reading and Interpreting Circuit Diagrams

Circuit diagrams are the blueprints of electrical systems. They offer a visual representation of how components like wires, resistors, and power sources are connected. Being able to read and interpret these diagrams is important for anyone working with off-grid solar power systems, as it allows you to understand how electricity flows and how various components interact.

The Basics of Circuit Diagrams

At first glance, a circuit diagram might seem overwhelming, filled with symbols, lines, and connections. But once you understand the basics, it becomes much easier to decode.

1. **Lines** – These represent the wires that connect different components in the circuit. In most diagrams, straight lines indicate a direct connection, while lines crossing each other without a connection are shown by a small arc or a break.

2. **Components** – Each component, whether it's a battery, resistor, or solar panel, is represented by a unique symbol. Understanding these symbols is a key to reading the diagram correctly.

3. **Flow of Current** – Circuit diagrams typically show the path that electricity takes as it flows from the power source, through various components, and back to the source. Arrows might be used to indicate the direction of current flow.

Interpreting a Circuit Diagram

Let's look at a simple example a basic circuit that illuminates an LED. The diagram depicts a battery linked to a resistor, which is subsequently connected to an LED. Interpreting the symbols and connections allows you to comprehend how the current goes from the battery, via the resistor (which restricts the current), and eventually through the LED, causing it to light up.

Common Electrical Symbols

To understand and construct circuit diagrams, you must be familiar with standard electrical symbols. These symbols are standardized for uniformity across diagrams, making electrical designs easier to comprehend and explain.

Here's a short reference for some of the most often used symbols:

Battery: Represents a power source with one or more cells.

Resistor: Indicates a component that resists the flow of current, used to control voltage and current in a circuit.

Capacitor: Stores electrical energy temporarily, usually represented by two parallel lines.

Diode: Allows current to flow in one direction only, often depicted as a triangle pointing towards a line.

LED (Light Emitting Diode): Similar to a diode symbol but with arrows indicating light emission.

Ground: Represents a connection to the earth or a common return path for current, essential for safety and circuit stability.

Switch: A device that can open or close a circuit, controlling the flow of electricity.

Memorizing these symbols is vital, so understands how they operate inside a circuit. As you analyze various diagrams, take notice of how these symbols are utilized and interact with one another.

Designing Basic Circuit Diagrams

Now that you've learned how to read circuit diagrams and recognize basic electrical symbols, it's time to construct your own. Whether you're designing a basic off-grid solar setup or a more complicated system, the ability to draw clear, precise schematics is very, very important.

Steps to Designing a Circuit Diagram

1. **Define the Objective**: What do you want your circuit to do? Whether it's lighting a light bulb or charging a battery, begin with a clear purpose.

2. **List the Components**: Determine the components you'll require based on your aim, such as solar panels, batteries, inverters, and controllers.

3. **Sketch the Layout**: Begin with a crude drawing, placing your power source (such as a battery) in the center. Next, add the other components, using

straight lines to represent cables. Make careful you use the right symbols for each component.

4. **Determine Connections**: Determine how each component will be linked. Will they be arranged in series or parallel? Ensure that the connections enable efficient and safe functioning.

5. **Review and Refine**: Once your diagram is finished, check it for correctness. Check that all components are accurately labeled and that the power flows logically and efficiently.

6. Simulate or test the circuit on a breadboard before finishing the design. This helps uncover any possible concerns prior to execution.

Practical Example: Designing a Simple Solar Charging Circuit

Let's imagine you want to create a simple circuit to charge a 12V battery with a solar panel. You would begin by drawing up your power source (solar panel) and batteries. Next, you'd place a solar charge controller between them to oversee the charging process and keep the battery from overcharging. Your circuit diagram would show the solar panel linked to the charge controller, followed by another connection from the charge controller to the battery. To make a legible and functioning circuit diagram, follow these steps and use the right symbols.

Understanding Energy and Power Consumption

Two essential ideas underpin all electrical systems: energy and power. Though commonly used interchangeably, they indicate different aspects of electricity, which are crucial to understand.

- **Energy** is the ability to perform labor, measured in watt-hours (Wh) or kilowatt-hours (kWh). It reflects the overall quantity of electricity utilized throughout the course of time.

- **Power** is the rate at which energy is used or created, expressed in watts (W) or kilowatts (kW).

It informs you how much energy is consumed or generated at any particular instant.

To visualize this, imagine energy as the total amount of water stored in a tank and power as the rate at which water pours from a tap. The faster the water flows (more power), the quicker the tank (energy) empties.

Understanding both of these notions for your off-grid system will assist you in determining how much energy your house requires on a daily basis and how powerful your solar system must be to supply that demand.

Calculating Energy Usage

To build an effective off-grid solar power system, you must first establish your average energy use. This involves calculating the energy consumption of each item or device that you intend to power.

Here's how to calculate your energy use:

1. **List any electrical appliances**: you'll use, including lights, refrigerators, fans, and computers.

2. **Check the Power Rating**: Each appliance has a power rating, usually found on a label or in the user manual, expressed in watts (W).

3. **Estimate Usage Time**: Determine how many hours per day each appliance will run. This is crucial for understanding your total energy consumption.

4. **Calculate Daily Energy Use**: Multiply the power rating by the number of hours used each day to get the daily energy consumption for each appliance. For example, if a 60W light bulb is used for 5 hours a day, it consumes 60W x 5 hours = 300Wh/day.

5. **Sum It Up**: Add up the energy consumption of all your appliances to get your total daily energy requirement.

Estimating Power Requirements for Off-Grid Systems

Once you've determined your daily energy demands, the following step is to calculate the power requirements for your off-grid system. This requires a few more considerations.

- **Peak Power Demand**: Identify the times of day when your power demand is highest. This is when multiple appliances are running simultaneously. Your system needs to be capable of handling this peak load.

- **Inverter Sizing**: The inverter converts DC power from your solar panels into AC power for your appliances. To size it correctly, consider the peak power demand. If your peak load is 2,000 watts, you'll need an inverter that can handle at least that much power.

- **Battery Storage**: Batteries store energy for use when the sun isn't shining. Estimate how many kilowatt-hours of storage you need by considering your daily energy usage and how long you might need to rely on stored energy (e.g., during cloudy days).

- **Solar Panel Sizing**: Finally, determine the number and size of solar panels required to generate enough energy to meet your daily consumption and charge your batteries. Consider the average sunlight hours in your location to ensure your panels can produce sufficient energy.

By carefully estimating your power requirements, you can design a solar power system that's well-suited to your energy needs, ensuring you have enough power without overspending on unnecessary capacity.

Managing Power Consumption for Efficiency

Efficient power management is an important fact to maximizing the performance and longevity of your off-grid system. Here are some tips to help you manage power consumption effectively.

- **Prioritize Essential Loads**: Focus on powering essential appliances like lights, refrigeration, and communication devices first. Luxury items like entertainment systems can be secondary priorities.

- **Energy-Efficient Appliances**: Invest in energy-efficient appliances. Look for devices with high energy efficiency ratings, such as LED lights, energy-star-rated refrigerators, and inverters with low standby power consumption.

- **Use Power Wisely**: Develop habits that reduce energy usage. Turn off appliances when not in use, use timers to control energy consumption, and avoid running multiple high-power appliances simultaneously.

- **Monitor and Adjust**: Regularly monitor your system's performance and energy usage. If you find that you're running out of power too quickly, look for ways to reduce consumption or consider expanding your system.

CHAPTER THREE

ESSENTIAL TOOLS & EQUIPMENT

Safety Gear and Precautions

Working with electrical systems may be dangerous. When connecting solar panels, installing batteries, or dealing with inverters, safety should always come first. Here's all you need to know to defend yourself and others around you.

Personal Protective Equipment (PPE)

When working with solar power systems, proper Personal Protective Equipment (PPE) is not only advised; it is needed. The right equipment may make the difference between a minor mishap and a serious injury. Here is what you should have on hand:

- **Insulated Gloves**: These are your principal means of protection from electrical shocks. Look for gloves rated for the voltage you'll be working with, and make sure they're in great shape before each use.

- **Safety Goggles**: Keep your eyes safe from flying debris, sparks, and unintended chemical spills. Clear, shatterproof goggles are necessary while cutting wires, drilling, or handling batteries.

- **Hard Hat**: If you're working outside, particularly while installing solar panels on a roof, a hard helmet will keep you safe from falling objects and mishaps.

- **Protective Footwear**: Steel-toed boots with rubber soles protect your feet from heavy falling objects and provide insulation against electrical currents.

- **High-Visibility Vest**: If you operate in a crowded area or near a road, wearing a high-visibility vest can help ensure that people notice you.

- **Long-Sleeved Clothing**: Wear long-sleeved flame-resistant clothing to avoid burns from hot surfaces or exposure to hazardous substances.

Always inspect your personal protective equipment (PPE) before use. If an item is broken or worn out, it should be replaced immediately. Your safety is definitely worth the investment.

Electrical Safety Guidelines

Although electricity is not visible, the threats it poses are real. Here are some important guidelines to remember while working with electrical components in your solar power system.

- **De-energize Before Working**: Always switch off the power when working on an electrical component. This cannot be negotiated. Before touching any wires or terminals, use a voltage tester to check that no current is flowing.

- **Understand Your Circuit**: Before making any changes or installations, make sure you fully understand the circuit you're working on. Know where the power source is, how it is connected, and where potential threats are.

- **Avoid Water**: Never work on an electrical system in wet weather. Water conducts electricity, and even a small amount of moisture can create a dangerous situation.

- **Use Insulated Tools**: When working with electrical systems, always use tools with insulated handle. This insulation can assist prevent accidental shocks and provide an additional layer of protection.

- **Work in Pairs**: When working with electrical systems, it's preferable to have someone around. In case of an accident, this person can phone for help and assist in an emergency.

Handling and Storing Tools Safely

Having the correct equipment is good, but understanding how to use and keep it properly is important. Here are some tips for keeping your tools and yourself safe.

- **Keep Tools Organized**: A messy office is a dangerous workplace. Keep your tools neatly stored and conveniently accessible. Organize your tools using toolboxes, pegboards, or tool belts.

- **Inspect Tools Regularly**: Before using any tool, check for damage. Look for frayed cords, broken handles, and other signs of wear and tear. Damaged tools should be fixed or replaced quickly.

- **Use Tools for Their Intended Purpose**: It's tempting to use a screwdriver as a chisel or a wrench as a hammer, but this might result in accidents or damage. Always utilize tools as intended.

- **Store Tools Properly**: After using your tools, put them in a dry, secure spot. Moisture may cause metal tools to rust, and incorrect storage can make them dull or ruined.

- **Lift with Care**: When carrying big tools or equipment, employ the right procedures. To prevent injury, bend your knees, keep your back straight, and lift using your legs rather than your back.

- **Disconnect Power Tools When Not in Use**: If you use power tools, keep them unplugged from the power source while not in use. This avoids inadvertent activation, which might cause damage.

Power Tools and Their Applications

Power tools are the foundation of every serious operation, and when it comes to solar installation, they are essential. These tools are intended to make your job easier, faster, and more exact, ensuring that every component of your solar system is securely and precisely fitted.

Cordless Drill & Impact Driver

A cordless drill is perhaps the most versatile equipment in your arsenal. Its primary use is to drill holes, but with the appropriate attachments, it may also be used to mix materials and drive screws. The beauty of a cordless drill is its mobility and power, which allows you to work in distant areas without the need for a power source.

An impact driver, on the other hand, is built for heavy-duty jobs. It has greater torque than a normal drill, so it's perfect for driving lengthy screws or bolts into resistant materials. When working on solar projects, particularly when installing panels or anchoring structural components, an impact driver ensures that fasteners are snugly and securely fastened.

Soldering Iron & Heat Gun

In any solar project, you'll frequently need to join wires or attach components that a simple twist and tape won't do. Here is when a soldering iron comes in handy. It enables you to build robust, lasting electrical connections, ensuring that your system is both efficient and safe over time. Proper soldering is essential for avoiding difficulties like voltage dips or faulty connections, which might jeopardize your entire system.

A heat gun complements the soldering iron by allowing you to put heat shrink tubing over connections, adding an extra layer of protection. It's also handy for jobs like removing adhesive bindings and bending PVC pipes into specific forms for installation.

Saw Types Jig Saw, Reciprocating Saw, and hole Saw

Cutting materials to the proper size and form is a necessary component of every solar installation. Different saws serve different purposes based on the material being sliced:

- **Jig Saw:** The jigsaw is a very versatile instrument that excels at producing accurate, curved cuts in materials such as plywood or plastic. It's especially useful when you need to cut unusual shapes or notches for your solar panels or components.

- **Reciprocating Saw:** A reciprocating saw is known for its strength and ability to cut through almost anything, making it excellent for demolition or rough cuts in dense materials such as metal and wood. Its powerful cutting motion makes it ideal for swiftly removing barriers or reducing structural pieces to match your solar system.

- **Hole Saw:** When drilling big, circular holes, such as for conduit or ventilation, a hole saw is the instrument to use. It is intended to cut through a range of materials, including metal, wood, and drywall, with accuracy and efficiency.

Screwdrivers and Their Types

Despite all of the power tools available, the basic screwdriver remains one of the most important items in your toolbox. Having the appropriate screwdriver may save you time and stress, since various jobs require different kinds.

- **Flathead Screwdriver:** This traditional tool is intended for screws with a single horizontal groove. It's good for ordinary activities, but it may also be used to pry open cases or push components into position.

- **Phillips Screwdriver:** The Phillips screwdriver's cross-shaped tip is ideal for screws built to withstand increased torque. This is especially beneficial in solar projects where secure communications are required.

- **Torx Screwdriver:** Torx screwdrivers, with their star-shaped tips, are becoming increasingly prominent in solar equipment. It's intended to prevent cam-out and allow for high-torque applications without damaging the screw head.

- **Precision Screwdrivers:** These little screwdrivers are useful for delicate activities like attaching wires to small terminals and adjusting sensitive components. Having a set of precision screwdrivers allows you to complete any little, complicated jobs that arise during installation.

Wiring Tools and Accessories

Wiring is the foundation of every solar power system, linking all components to ensure a continuous flow of electricity. A collection of specialist wiring tools and accessories will be required to ensure the seamless operation of your system. These tools not only simplify your task, but also assure that the connections you establish are safe, secure, and long-lasting.

Wire Cutters & Strippers

Wire Cutters are essential for cutting wires to the proper length. A clean, exact cut is required to avoid frayed ends, which can cause weak connections or short circuits. Wire strippers, on the other hand, are used to remove insulation from wire ends while leaving the metal intact. Stripping wires correctly is essential for creating secure connections, whether you're crimping connectors or putting them into terminals.

Imagine attempting to join wires without these tools—it would be like writing without a pen. Quality wire cutters and strippers will make your job neater, faster, and safer, ensuring that each connection is as dependable as the next.

Crimping Tools and Connectors

Crimping Tools are used to attach connectors to the ends of cables, which is also important for making a stable connection in your solar system. Crimping is the process of squeezing a metal connection onto a wire, resulting in a tight attachment that is difficult to loosen. This is especially significant in solar installations, where steady electrical supply is required.

Without the right crimping equipment, you risk having weak connections, which can lead to electrical failures or even fires. Quality connectors combined with the appropriate crimping equipment guarantee that your connections are sturdy and dependable, able to survive the weather and the passage of time.

Multi-meter and Clamp Meter

A Multi-meter is one of the most useful instruments in your solar toolbox. It detects voltage, current, and resistance, allowing you to troubleshoot and ensure that your system is running properly. Whether you're evaluating the output of your solar panels or monitoring the charge of your batteries, a multi-meter gives the information you need to keep everything working properly.

A Clamp Meter is very handy for monitoring current without disconnecting the wire. This instrument can monitor the flow of electricity via a conductor by simply clamping it around the wire, making it perfect for routine maintenance and rapid diagnostics.

Think of these instruments as solar system health monitors. They help you keep an eye on its vital indicators, ensuring everything is running safely and efficiently.

Measuring Tape & Fish Tape

Precision is essential while installing your solar power system, which is where a Measuring Tape comes in. Accurate measurements are required for everything from installing solar panels to cutting cables to the proper length. A reliable measuring tape guarantees that your system components are properly positioned and that cables reach their destinations without being too short or too lengthy.

Fish Tape is another useful instrument, especially for pulling wires through conduits. It's a flexible, flat wire that guides electrical lines into narrow areas, making the wiring procedure considerably easier and less time-consuming.

These tools may appear basic, but they play an important role in ensuring that your installation is clean, efficient, and professional.

Junction Boxes & Bus-bars

Junction Boxes are used to keep electrical connections safe from the environment and avoid inadvertent contact.

They are vital for arranging your wiring and creating a safe environment for your electrical connections, proper use of junction boxes can assist prevent shorts and extend the life of your system.

Bus-bars are metal strips or bars that provide electricity to many circuits in your system. They provide a clean, orderly method for connecting several wires to a single location, ensuring that electricity is transmitted evenly and effectively. Using bus-bars decreases the complexity of wiring, minimizes the danger of mistakes, and contributes to a neat and safe electrical setup.

These components may be hidden after your system is setup, but they are the unsung heroes who keep everything functioning smoothly behind the scenes. Without them, your system will be chaotic, prone to failure, and possibly harmful.

Solar Racking and Mounting Equipment

The racking and mounting system is a vital component of every solar installation. This is the framework that keeps your solar panels in place, whether on your roof, the ground, or another building. Choosing the proper mounting equipment is vital for the durability and effectiveness of your system.

Rails and Roof-Mount Flashings

Rails are the core of any racking system, serving as a solid base for your solar panels. They are often composed of robust materials such as aluminum, which is both strong and lightweight, allowing for simple installation while guaranteeing that the panels remain firmly in place over time.

Roof-mount systems sometimes require flashings, which are intended to keep water from leaking into your roof where the mounts penetrate. Flashings are essential for preserving the integrity of your roof and protecting it from leaks and the elements. When placed correctly, they provide a watertight seal around the mounting points, guaranteeing that your solar system does not jeopardize the safety of your home.

Consider a roof without correct flashings; over time, water seeps through, creating undetected damage. Months later, you are confronted with a pricey repair. That's why excellent flashings are not just a suggestion, but a need.

End Clamp & Mid Clamp

End clamps and mid clamps are modest but important components that hold the solar panels to the rails. End clamps are utilized at the array's borders to securely fasten the outer panels. Mid clamps, on the other hand, are put between each panel to ensure that they are securely attached to the rail system with no gaps.

These clamps must be properly positioned and tightened to keep the panels fixed even in high winds or thick snow. A loose clamp may cause panels to move or even detach over time, which is not only dangerous but also reduces the effectiveness of your system.

Consider the clamps the unsung heroes of your installation. Though modest, their duty is monumental: keeping your investment secure and operational.

Tilt Mounts and Ground Mounts

Not all solar systems are roof-mounted. Tilt mounts or ground mounts may be quite useful in some situations, particularly when roof space is limited or the roof is not properly angled.

Tilt mounts are used to place solar panels on flat surfaces like a roof or the ground. These attachments enable the panels to be positioned properly to collect the maximum sunlight, usually at an angle proportional to your latitude. Adjusting the tilt allows you to maximize energy output, particularly during the winter months when the sun is lower in the sky.

Ground mounts are another option that is frequently employed in situations when roof installation is neither practical nor desirable. These methods allow for bigger solar arrays and may be deployed in open areas with enough sunlight exposure. Ground mounts are firmly secured into the ground and, like tilt mounts, may be adjusted for the ideal angle to catch sunlight.

Site-specific considerations frequently influence the decision between roof mounts, tilt mounts, and ground mounts. For example, if your roof is shadowed by trees or other structures, ground mounts may be a better choice. Alternatively, in areas with severe snowfall, tilt mounts can be modified to minimize snow accumulation on the panels, which would otherwise reduce efficiency.

Battery Maintenance Tools

Your solar system's batteries are the core of your off-grid arrangement, storing the energy generated by your panels for later usage when the sun isn't shining. Proper battery maintenance is crucial for extending their life and keeping your system running properly. The tools in this area will let you monitor and maintain your batteries precisely.

Hydrometer & Refract meter

Understanding the condition of your batteries is necessary to their longevity and performance. A hydrometer and a refract meter are useful in this situation.

- **Hydrometer**: A hydrometer monitors the specific gravity of the electrolyte in your batteries, providing a clear indication of their level of charge. By pulling a little sample of electrolyte into the hydrometer, you can rapidly determine if your batteries are fully charged, partially charged, or in need of repair. Regular usage of a hydrometer helps avoid overcharging or deep discharging, both of which can seriously harm your batteries.

- **Refract meter**: While comparable to a hydrometer, a refract meter is more accurate and user-friendly, particularly in colder locations. It uses the refractive index of the battery's electrolyte to calculate its specific gravity. This gadget is especially handy if you need very accurate readings or if you're working with various sorts of batteries that require more careful monitoring.

Both tools are required for anyone serious about maintaining their solar battery setup. Regularly monitoring your batteries using a hydrometer or refract meter allows you to detect any problems early on and take steps to maintain your system functioning at full performance.

Battery Monitoring Systems

A battery monitoring system is a must-have tool that gives real-time information on your battery bank's status.

A good battery monitoring system tracks an important data like as.

- **State of Charge (SOC)**: Shows how much energy remains in your battery bank.

- **Voltage and Current**: Monitors the flow of energy into and out of your batteries.

- **Temperature**: Tracks battery temperature to prevent overheating and damage.

With this information at your fingertips, you can improve your system's performance, minimize overcharging, and never be caught off guard by a low battery bank. Investing in a reliable battery monitoring system is one of the most effective methods to safeguard your solar power investment and retain energy independence.

Cleaning and Maintenance Supplies

Maintaining your solar power system entails more than simply monitoring; it also includes maintaining your equipment clean and in excellent shape. Over time, dust, filth, and corrosion can accumulate on your batteries, connectors, and other components, causing inefficiency or even system failure. The following supplies are required for routine maintenance:

- **Terminal Brushes**: Battery terminals are prone to corrosion, which can disrupt the flow of energy and degrade system performance. A terminal brush is a basic instrument used to remove rust off battery terminals, resulting in a firm, conductive connection. Brushing your terminals on a regular basis ensures that your system runs smoothly.

- **Cable Ties**: Maintaining wire organization is critical in any solar setup. Loose or twisted wires might result in inadvertent harm or ineffective functioning. Cable ties allow you to secure and organize cables, lowering the danger of wear and tear and making your system easier to monitor and repair.

- **Cleaning Supplies**: Dust and debris can accumulate on your solar panels, limiting their effectiveness, as well as on your batteries and other components. A package of cleaning materials, such as a soft brush, cloth, and non-corrosive cleanser, guarantees that all aspects of your system are clean and efficient. Regularly cleaning your panels, terminals, and other equipment extends their longevity and maintains maximum functioning.

CHAPTER FOUR

SOLAR PANEL SELECTION & INSTALLATION

Choosing the appropriate solar panels is an important step in your off-grid solar power adventure. With so many alternatives available, it's critical to grasp the differences and choose the best one for your individual requirements. This chapter will walk you through the choosing process and provide you complete instructions for installing your solar panels.

Types of Solar Panels and Their Efficiency

The efficiency of a solar panel is how well it turns sunlight into energy. Different solar panels have variable efficiency, prices, and compatibility for different situations. Below, we'll discuss the four major types of solar panels utilized in off-grid installations.

1. Mono crystalline Panels

Mono crystalline solar panels are noted for their high efficiency and streamlined design. These panels are constructed with a single, continuous crystal structure, allowing for higher energy conversion rates.

- **Efficiency:** Typically varies between 17% and 22%, making them one of the most efficient forms of solar panels available.

- **Advantages:** High efficiency, extended life, and exceptional performance in low-light settings.

- **Disadvantages:** Higher cost than other types of panels.

- **Best Use:** Ideal for installations with limited area and high efficiency requirements, such as roofs or urban locations.

2. Polycrystalline Panels

Polycrystalline panels are composed of several silicon crystals fused together. These panels are somewhat less efficient than mono crystalline panels, but offer a more inexpensive alternative.

- **Efficiency:** Usually ranges from 15% to 17%.

- **Advantages:** Lower cost, decent efficiency, and less waste produced during manufacturing.

- **Disadvantages:** Slightly lower efficiency and more prominent blue hue compared to mono crystalline panels.

- **Best Use:** Suitable for larger installations where space is not a primary concern, such as in rural or off-grid locations.

3. Thin-Film Panels

Thin-film solar panels are made by placing one or more layers of photovoltaic material onto a substrate. These panels are lightweight and flexible, making them versatile for various applications.

- **Efficiency:** Typically ranges from 10% to 12%, making them the least efficient of the three main types.

- **Advantages:** Lightweight, flexible, and can be applied to surfaces where traditional panels wouldn't fit, such as curved roofs or mobile installations.

- **Disadvantages:** Lower efficiency and shorter lifespan compared to crystalline panels.

- **Best Use:** Ideal for portable solar solutions, low-load applications, or installations where weight is a significant factor.

4. Bifacial Panels

Bifacial solar panels can capture sunlight from both sides, increasing their overall energy production. These panels are made with transparent back sheets or dual glass, allowing light to pass through and be reflected from the ground or other surfaces.

- **Efficiency:** Can exceed 22% under optimal conditions, depending on the reflectivity of the installation surface.

- **Advantages:** Higher overall energy production, especially in snowy or reflective environments.

- **Disadvantages:** Higher cost and require specific mounting setups to maximize reflected light capture.

- **Best Use:** Ideal for installations where there is high ground reflectivity, such as in snowy regions, or where maximizing energy output is important.

Installation Process

Once you've selected the appropriate solar panels for your needs, the next step is installation. Here's a summary of the process.

1. **Site Assessment:** Begin by assessing your installation site to establish the optimal position for your panels. Consider aspects such as sunshine exposure, shade, and structure integrity.

2. **Mounting System Installation:** Install the mounting hardware following the manufacturer's

directions. Make sure the mounts are tight and correctly oriented to optimize the angle of your panels for optimum solar exposure.

3. **Panel Placement:** Carefully install your solar panels on the mounting system. Secure them with the necessary hardware, ensuring that they are firmly attached and placed appropriately.

4. **Wiring:** Connect the panels with MC4 connectors, making sure that all connections are safe and weather-proof. Route the cable neatly to the charging controller and inverter.

5. **Final Checks:** Once your panels have been fitted and connected, run a series of tests to confirm that everything is working properly. Use your multimeter to verify the voltage output and then inspect all connections for stability.

Understanding Solar Panel Performance Curves

Solar panels are more than simply a collection of photovoltaic cells; they are precisely calibrated systems with distinct performance characteristics. Understanding the performance curves of your solar panels is vital for getting the most out of them. These curves show how effectively your panels will perform under various situations and help you maximize the electricity you capture from the sun.

I-V Curve

The I-V curve, also known as the Current-Voltage curve, is a valuable tool for understanding how your solar panel works. It depicts the relationship between the current (I) produced by the solar panel and the voltage (V) applied to it at any particular time. This curve demonstrates how your solar panel responds under various degrees of sunshine and load circumstances. The main lesson is that the form of the I-V curve helps you identify the maximum power point, which is the point where your panel produces the most power.

P-V Curve

The P-V curve (Power-Voltage curve) is closely connected to the I-V curve. It shows how the power output of your solar panel fluctuates with voltage. The apex of the P-V curve represents the Maximum Power position (MPP), the best position at which your solar panel produces the most power output. Understanding the P-V curve is critical for ensuring that your solar system runs at the most efficiency, especially when conditions are not optimal.

Maximum Power Point Tracking (MPPT)

Maximum Power Point Tracking (MPPT) is a technology used in current solar charge controllers to continually monitor and modify your solar panels' operating point so that it is as near to the MPP as feasible. MPPT guarantees that your panels constantly operate at peak efficiency, even when sunlight levels shift throughout the day. This technique is very vital for off-grid systems where optimal energy capture is needed for dependable power delivery.

Temperature Coefficients

Temperature significantly affects solar panel performance. The temperature coefficient describes how much a panel's power output reduces with each degree Celsius increase in temperature. High temperatures can reduce the voltage produced by solar cells, resulting in decreased total power production. Understanding the temperature coefficient of solar panels is critical, particularly if you reside in a hot region. Panels with lower temperature coefficients are desirable in such settings because they lose less efficiency when the mercury rises.

Effects of Temperature and Shading on Panels

Temperature and shadowing are two of the most prevalent issues that might reduce the efficiency of your solar panels. In this part, we'll look at how these components affect your system and what you can do to mitigate their consequences.

Impact of High Temperatures

High temperatures can greatly affect the effectiveness of your solar panels. As previously stated the voltage output of solar cells decreases as temperature increases, this implies that on especially hot days, solar panels may produce less electricity than they would in colder temperatures. However, this does not imply that solar panels are inefficient in hot climates rather it underscores the need of selecting panels with acceptable temperature coefficients and mounting them in a way that allows for sufficient ventilation to disperse heat.

Partial Shading and Its Effects

Shading is the adversary of solar panels. Even a minor amount of shadowing on a portion of your panel can significantly limit its overall power output. This is because solar cells on a panel are connected in series, and the cell that receives the least amount of light determines the overall performance of the string of cells. When a portion of the panel is darkened, it forms a bottleneck, limiting the current that can travel through the circuit. Understanding how to prevent or limit the impacts of shadowing is critical for increasing the effectiveness of your solar system.

Mitigating Shading Issues

There are various ways you may use to offset the impact of shade on your solar panels:

- **Optimal Placement:** Carefully plan the placement of your solar panels to avoid shading from trees, buildings, or other obstructions, especially during peak sunlight hours.

- **Panel Layout:** Consider the layout of your panels. Sometimes, arranging them differently can reduce the impact of shading.

- **Bypass Diodes**: Many modern solar panels use bypass diodes, which enable current to skip shaded cells, decreasing power loss. Ensure your panels have these diodes if shading is a concern.

- **Micro inverters or Power Optimizers:** Using micro inverters or power optimizers on each panel can help minimize the impact of shading by allowing each panel to operate independently at its maximum potential.

Hotspots and Their Prevention

Hotspots form when a darkened or damaged area of a solar panel warms up much faster than the remainder of the panel. These hotspots can cause long-term harm to the panel, decreasing its efficiency and longevity. To prevent hotspots:

- **Regular Maintenance:** Regularly inspect your panels for dirt, debris, or any physical damage that could cause shading or cell mismatch.

- **Proper Installation:** Ensure panels are installed with enough clearance to allow for air circulation, which helps dissipate heat and prevents localized overheating.

- **Monitoring:** Use a monitoring system to keep track of each panel's performance. This can help you detect potential hotspots early before they cause significant damage.

Optimal Tilt and Azimuth Angles

The tilt angle of your solar panels is more than just a technical detail; it's the key to harnessing the full potential of the sun's energy. The tilt angle refers to the angle at which the panels are inclined relative to the ground. The goal is to position your panels so that they receive the maximum possible sunlight throughout the year.

Calculating Tilt Angle for Maximum Efficiency

To calculate the optimal tilt angle, you need to consider your geographical location. A general rule of thumb is to set the tilt angle equal to your latitude. For example, if you live at 35 degrees latitude, your panels should be tilted at 35 degrees. However, this is just a starting point. Seasonal variations in the sun's position mean that adjusting the tilt angle periodically can further optimize energy production.

Azimuth Angle and Its Importance

While the tilt angle determines how high your panels are angled, the azimuth angle determines their horizontal orientation. The azimuth angle is measured clockwise from true north and is critical in ensuring your panels face the sun throughout the day.

In the Northern Hemisphere, the optimal azimuth angle is typically due south (180 degrees). This orientation maximizes sunlight exposure during peak hours. However, if your installation site doesn't allow for a perfect south-facing setup, slight adjustments can still yield good results. For instance, an azimuth angle within 20 degrees east or west of true south can still capture a significant amount of solar energy.

Seasonal Adjustments

The sun's path across the sky changes with the seasons, affecting how much sunlight your panels receive. To maximize efficiency, you can adjust your panels' tilt angle seasonally:

- **Winter:** Increase the tilt angle by 10-15 degrees more than your latitude to capture the lower winter sun.

- **Summer:** Decrease the tilt angle by 10-15 degrees less than your latitude to take advantage of the higher summer sun.

These adjustments ensure that your panels receive optimal sunlight throughout the year, increasing your system's overall efficiency.

Tilt Angle for Different Roof Types

Different roof types present unique challenges when it comes to setting the optimal tilt angle. For instance:

- **Flat Roofs:** Flat roofs offer flexibility, allowing you to set your panels at the optimal tilt angle using mounting structures.

- **Pitched Roofs:** The tilt angle on a pitched roof is often fixed by the roof's slope. If this angle is close to your optimal tilt angle, you're in luck. Otherwise, you may need to consider adjustments or use specialized mounting brackets.

- **Metal Roofs:** These roofs often come with their own challenges, such as dealing with standing seams or corrugations. However, with the right mounting systems, you can still achieve an efficient tilt angle.

Panel Mounting Techniques for Different Roof Types

Mounting solar panels correctly is crucial for their longevity and efficiency. Each roof type requires specific mounting techniques to ensure that the panels are securely fastened and optimally positioned.

Flat Roof Mounting

Flat roofs provide a blank canvas for solar installations, offering flexibility in panel orientation and tilt angle. Panels on flat roofs are usually mounted using adjustable racks that allow you to set the optimal tilt angle. It's important to consider the weight and wind load on the panels, ensuring the mounting structure is robust and secure.

Ballasted Mounts are a common choice for flat roofs, where the panels are held in place by heavy weights rather than being drilled into the roof. This method minimizes roof penetration and potential leaks but requires careful planning to ensure stability.

Pitched Roof Mounting

On a pitched roof, the panels are typically mounted parallel to the roof's slope. If the roof's pitch is close to the optimal tilt angle for your location, this setup is ideal. If not, you may need to use tilt-up brackets to adjust the angle.

Rail-based Mounting Systems are popular for pitched roofs, where rails are attached to the roof with hooks or brackets, and the panels are then secured to these rails. This method allows for slight adjustments to the panel angle and ensures a strong, wind-resistant installation.

Metal Roof Mounting

Metal roofs, especially those with standing seams, require specialized mounting solutions. The key is to avoid penetrating the metal, which can lead to leaks.

Clamping Systems are commonly used for metal roofs, where clamps attach directly to the standing seams, holding the panels securely in place without drilling into the roof. This method is both effective and minimizes the risk of water ingress.

Ground Mounting Options

If roof space is limited or unsuitable, ground mounting provides an excellent alternative. Ground-mounted systems offer the freedom to position your panels in an ideal location and at the optimal tilt and azimuth angles.

Pole Mounts are popular for small systems, where panels are mounted on a single pole that can be manually adjusted for tilt. For larger installations, **Rack Mounts** allow multiple panels to be installed in a row, often with automated tracking systems that adjust the panels throughout the day to follow the sun.

Ground mounting also allows for easier maintenance, as the panels are more accessible. However, it requires adequate space and careful site selection to avoid shading from trees or buildings.

Panel Placement and Spacing

The placement of your solar panels is crucial to maximizing energy production. Start by identifying the sunniest spots on your property—areas that receive the most direct sunlight throughout the day. Ideally, these locations should face south in the Northern Hemisphere or north in the Southern Hemisphere to capture the maximum amount of sunlight.

Panel Spacing is another key factor. Your panels need to be spaced adequately to prevent shading from each other and to allow for sufficient air circulation. Proper airflow beneath and around the panels is essential to prevent overheating, which can decrease their efficiency. Ensure that there is enough room for maintenance and cleaning, as well as to avoid interference from other structures or natural elements.

Avoiding Shading from Nearby Objects

Shading is the enemy of solar panel efficiency. Even a small amount of shade on a single panel can significantly reduce the overall output of your entire system. It's important to assess potential shading from nearby trees, buildings, or other obstructions. As the sun's position changes throughout the day and year, ensure that your panels will remain un-shaded, especially during peak sunlight hours. Trimming trees and positioning panels away from potential obstructions can help mitigate shading issues.

Panel Spacing for Air Circulation

Air circulation is critical for maintaining the performance and longevity of your solar panels. When installing on a roof, ensure there is a gap between the panels and the roof surface to allow air to flow underneath. This ventilation helps dissipate heat and prevents the panels from overheating, which can lead to a drop in efficiency. Ground-mounted systems should be installed at a height that allows for similar airflow.

Considerations for Roof Load

Before installing solar panels on your roof, it's important to assess whether your roof can handle the additional load. Solar panels, mounting systems, and the weight of accumulated snow (in colder climates) can add significant stress to your roof. Consult with a structural engineer or a professional installer to evaluate your roof's load-bearing capacity and determine if any reinforcements are necessary.

Connecting Solar Panels

Connecting solar panels correctly is vital for the safety and efficiency of your system. The way you connect your panels affects the voltage and current output, which in turn influences the performance of your entire setup.

Series vs. Parallel Connections are the two primary methods for wiring solar panels.

- **Series Connections**: In a series connection, the positive terminal of one panel is connected to the negative terminal of the next, and so on. This setup increases the voltage while keeping the current the same as that of a single panel. Series connections are useful when you need to match the voltage of your solar array with your inverter or charge controller.

- **Parallel Connections**: In a parallel connection, all the positive terminals are connected together, and all the negative terminals are connected together. This configuration increases the current while keeping the voltage the same as that of a single panel. Parallel connections are beneficial when you need to increase the current output to match the requirements of your batteries or other system components.

Designing the Panel Array

When planning your panel array, think about both the physical arrangement and the electrical setup. The physical arrangement should optimize solar exposure while still fitting within the available area, whether on a roof or on the ground. The electrical configuration whether series parallel or a combination of the two must meet the voltage and current needs of your system components.

A well-designed panel array balances the need for high efficiency with practical considerations like space, shading, and installation complexity. Take the time to plan your array carefully, as this will impact the overall performance and reliability of your system.

Using MC4 Connectors

MC4 connectors are the industry standard for solar panel connections, renowned for their dependability and simplicity of use. These connections enable you to easily and securely connect and disconnect panels, making installation and maintenance easier. Ensure that all connections are secure, waterproof, and meet your system's unique needs. The proper use of MC4 connectors decreases the possibility of loose connections, which can lead to electrical problems or system inefficiencies.

Series-Parallel Configurations

For bigger solar arrays, a series-parallel configuration—which combines series and parallel connections—is frequently the best option. This configuration allows you to reach the appropriate voltage and current levels by combining the advantages of series and parallel circuits. For example, you might join groups of panels in series to enhance voltage then connect those groups in parallel to boost current. This adaptable setup can assist enhance the performance of your system based on your individual energy requirements and the characteristics of your solar components.

Solar Panel Installation Checklist

Before you begin the installation procedure, make sure you have everything in place. Use this checklist to keep track of the important factors necessary for a successful solar panel installation:

- **Site Assessment**: Evaluate the location for optimal sunlight exposure and structural suitability.

- **Panel Selection**: Choose panels that match your energy needs, budget, and environmental conditions.

- **Mounting System**: Ensure you have a robust mounting system appropriate for your roof type or ground setup.

- **Permits and Regulations**: Obtain all necessary permits and adhere to local regulations and building codes.

- **Wiring and Electrical Components**: Prepare the necessary wiring, connectors, and protection devices for a safe and efficient system.

- **Safety Gear**: Equip yourself with the necessary safety gear, including gloves, goggles, and a harness for roof work.

Tools and Equipment Needed

A successful installation requires the right tools. Here's what you'll need:

- **Drill and Drill Bits**: For securing mounting brackets and panels.

- **Wrench Set**: For tightening bolts and nuts.

- **Screwdriver Set**: Both flathead and Phillips for various screws.

- **Wire Stripper and Cutter**: For preparing and managing wiring.

- **Use a multimeter** to evaluate electrical connections for correct voltage.

- **Level**: To ensure your panels are mounted correctly for maximum efficiency.

- **Ladder**: For safe access to your roof or elevated installation site.

- **Safety Harness**: Essential for working at heights.

Safety Precautions

Safety should always be your top priority during installation. Here are some key precautions.

- **Work in Pairs**: Never work alone, especially when working at heights.

- **Secure the Area**: Ensure the installation area is clear of obstructions and that your ladder is stable.

- **Wear Protective Gear**: Use gloves, goggles, and a helmet to protect against potential injuries.

- **Check Weather Conditions**: Avoid installation during wet, windy, or extreme weather conditions.

- **Disconnect Power**: Ensure that all electrical circuits are de-energized before you begin wiring.

Installation Steps

Installing your solar panels involves a series of careful, deliberate actions. Here's how to proceed.

1. **Mark the Installation Site**: Use a level and chalk to mark where the mounting brackets will be placed.

2. **Install Mounting Brackets**: Secure the brackets to your roof or ground mounts, ensuring they are level and properly spaced according to your panel specifications.

3. **Attach the Solar Panels**: Carefully position and secure each panel onto the mounting brackets, tightening bolts and screws firmly.

4. **Wiring the Panels**: Connect the panels according to your system's wiring diagram, ensuring all connections are secure and weatherproofed.

5. **Connect to the Inverter**: Run the wiring from the panels to your solar inverter, ensuring correct polarity and connections.

6. **Grounding**: Properly ground your system to prevent electrical hazards.

7. **Power Up**: After a final check of all connections and components, switch on your system and monitor the output.

Final Inspection and Testing

Once your panels are installed, it's time for a thorough inspection and testing:

- **Check Connections**: Verify all electrical connections are secure and free from corrosion or damage.

- **Test Voltage and Current**: Use a multi-meter to ensure your panels are producing the expected voltage and current.

- **Structural Integrity**: Inspect the mounting system for stability and security.

- **Monitor System Performance**: Over the first few days, monitor the system's performance to ensure everything is working as expected.

Maintenance and Cleaning of Solar Panels

Regular maintenance is essential to keep your solar panels operating at peak efficiency. Here's how to ensure your panels remain in top condition:

Routine Maintenance Tasks

- **Visual Inspection**: Check your panels regularly for any signs of damage, debris, or loose connections.

- **Monitor Performance**: Keep an eye on your system's performance metrics to spot any unusual drops in efficiency.

- **Check for Obstructions**: Ensure that no new objects, such as tree branches, are shading your panels.

Cleaning Techniques and Frequency

- **Cleaning Frequency**: Depending on your location and the amount of dust or debris in the air, clean your panels every 6-12 months.

- **Cleaning Method**: Use a soft brush or squeegee with a mixture of water and mild soap. Avoid utilizing strong chemicals and abrasive materials.

- **Rinse and Dry**: After scrubbing, rinse the panels with clean water and allow them to air dry. Avoid cleaning on hot, sunny days to prevent streaks.

Troubleshooting Common Issues

Even with careful installation and maintenance, issues can arise. Here's how to troubleshoot common problems:

- **Reduced Power Output**: Check for shading, dirt accumulation, or damaged panels.

- **Faulty Wiring**: Inspect all wiring for loose connections or wear and tear.

- **Inverter Issues**: If your inverter isn't working properly, check for error codes and refer to the manufacturer's troubleshooting guide.

- **Battery Problems**: For systems with battery storage, monitor the battery's charge level and health. Replace if necessary.

Warranty Types and Coverage

Solar panel warranties are your safety net, offering protection against unexpected issues that may arise with your system. Generally, there are two main types of warranties to be aware of:

1. Performance Warranty:

This warranty guarantees that your solar panels will produce a certain percentage of their original capacity over a specified period, typically 25 years. For example, a panel might come with a performance warranty stating that it will still produce 80% of its original output after 25 years. This type of warranty ensures that your panels will continue to perform efficiently over the long haul.

2. Product (or Equipment) Warranty

This warranty covers any defects in materials or workmanship that might occur during the manufacturing process. The duration of a product warranty can vary, but it typically lasts between 10 to 25 years. The duration of a product warranty can vary, but it typically lasts between 10 to 25 years. If your panels fail due to a manufacturing defect within this period, the manufacturer will repair or replace them at no cost to you.

Understanding these warranties is a key to making an informed decision when selecting solar panels. Look for panels that offer both robust performance and product warranties to ensure your system's longevity and reliability.

Factors Affecting Panel Lifespan

The lifespan of solar panels is influenced by a variety of factors, which can impact their efficiency and overall performance over time.

1. Quality of Materials

A higher-quality material generally leads to longer-lasting panels. When selecting solar panels, consider those made with premium materials, as they are less likely to degrade quickly and more likely to stand up to environmental challenges.

2. Installation Quality

Proper installation is critical to the longevity of your solar panels. Panels that are installed incorrectly are more prone to damage and inefficiency. Always work with certified professionals who have a proven track record in solar installations.

3. Environmental Conditions

The environment plays a significant role in the wear and tear on solar panels. Extreme weather conditions, such as heavy snow, high winds, or intense heat, can all affect the lifespan of your panels. Panels designed to withstand specific environmental factors are better suited for areas with harsh climates.

4. Maintenance and Care

Like any other investment, solar panels require regular maintenance to perform optimally. Cleaning the panels periodically to remove dirt, dust and debris as well as inspecting them for damage, can help extend their lifespan.

5. Degradation Rate

All the solar panels degrade over time, meaning they gradually produce less electricity as they age. However, the pace of degradation might vary. Panels with a lower degradation rate are preferable, as they will continue to produce more energy over a longer period.

What to Anticipate from Your Solar Panels over Time

Over the years, your solar panels will experience some degree of performance decline, but this is a normal part of their lifecycle. Here's what you can generally expect.

1. Initial Years (0-5 years)

In the first few years, your solar panels will operate at peak efficiency, often producing more electricity than you initially anticipated. This is the honeymoon period for your system, where everything is new and working optimally.

2. Midlife (5-15 years)

As your system enters its middle years, you may start to see a slight decline in output due to natural degradation. However, this decrease is usually minimal, and your panels will continue to generate a significant amount of electricity.

3. Later Years (15-25+ years)

In the latter part of their lifespan, your solar panels will likely experience more noticeable declines in efficiency. By the end of their warranty period, they may be operating at around 80% of their original capacity. While this is lower than when they were new, it still represents a considerable amount of energy production.

Despite this gradual decline, well-maintained panels can continue to produce electricity beyond their warranty period. Some panels may even last 30 years or more, providing clean energy and savings for decades.

Next

CHAPTER FIVE

SOLAR CHARGE CONTROLLERS

Solar charge controllers are frequently the hidden heroes in off-grid solar power systems. They may not be as visible as solar panels or as powerful as inverters, but their job is critical to your system's efficiency and lifetime. This chapter will go deeper into the operation and relevance of solar charge controllers, describing how they protect your batteries, manage power flow, and control loads to ensure your system runs smoothly.

Functions of a Solar Charge Controller

The solar charge controller, which is at the core of any off-grid solar power system, serves a variety of functions that are important to the health and operation of your system. Let us go out the key functions.

Battery Protection

One of the most important functions of a solar charge controller is to prevent your batteries from overcharging and severe discharge, both of which can dramatically limit their life. The controller checks the battery's charge level and regulates the flow of power from the solar panels to prevent overcharging. When the battery is fully charged, the controller decreases or stops the flow of current, keeping the battery in ideal condition.

Furthermore, if the battery's charge falls below a set level, the controller can turn off the power supply to avoid severe discharge, which can harm the battery and lower its capacity over time. This protective feature not only increases the life of your batteries, but also guarantees that your solar power system is dependable and efficient.

Power Regulation

Solar charge controllers regulate the power flow from solar panels to batteries and loads. Without adequate control, the fluctuating nature of solar energy due to changing sunshine conditions may result in ineffective charging or even damage to batteries and linked equipment.

The controller smoothies out these variations resulting in a steady and regulated flow of power, this management is especially critical when the solar panels generate more power than the batteries can hold or when the power consumption of linked devices varies. By controlling power, the controller maximizes energy collection and ensures system stability.

Load Control

Solar charge controllers frequently provide load control features. This feature enables the controller to regulate the power distribution to connected devices, ensuring that vital loads are prioritized and the battery is not exhausted prematurely.

Load control can also include unplugging non-essential loads when the battery charge is low to save power for critical systems. This capability is especially beneficial in off-grid configurations where power management is critical to ensuring that key equipment stay operational, even during periods of low sunshine.

Overview of Charge Controller Roles

Solar charge controllers are the brains of your off-grid solar system, continually monitoring, regulating, and optimizing the flow of energy to safeguard and improve its performance. Here's a short outline of the main functions they play.

- **Voltage Regulation:** Ensures that the voltage from the solar panels is properly matched to the battery bank's requirements, preventing overvoltage or under voltage conditions.

- **Current Control:** Limits the amount of current flowing into the batteries to avoid overcharging and potential damage.

- **Power Optimization:** Enhances the efficiency of your solar power system by managing energy flow based on current solar production and energy demands.

- **System Safety:** Provides a layer of safety by preventing conditions that could lead to equipment failure or unsafe operating conditions.

There are several types of solar charge controllers, including PWM and MPPT controllers. Understanding their roles and choosing the appropriate kind for your system is critical to obtaining a balanced and effective solar setup.

To summarize, a solar charge controller is an essential component of your off-grid solar system. It protects your batteries, regulates the power, and manages your loads effectively. By performing these vital functions, the charge controller allows you to get the most out of your solar power system, increasing battery life and improving overall system performance.

PWM vs. MPPT Charge Controllers

There are two types of solar charge controllers: pulse width modulation (PWM) and maximum power point tracking (MPPT). Understanding the distinctions between these technologies is critical for maximizing your solar system's performance and ensuring it runs at optimal efficiency.

What is PWM (Pulse Width Modulation)?

PWM is a technology that has been around for decades. It works by gradually lowering the power from the solar panels as the battery nears full charge, preventing overcharging and increasing battery life. PWM controllers are easier to construct and less expensive than their MPPT equivalents, making them a popular choice for smaller or budget-conscious solar arrays.

Advantages of PWM Charge Controllers

- **Cost-Effective:** PWM controllers are often less expensive, making them a perfect alternative for individuals who are just getting started with solar power or have a limited budget.

- **Simplicity:** PWM controllers are easy to install and maintain since they use fewer components and have a simple design.

- **Durability:** Because of its simplicity, PWM controllers are frequently more robust and dependable, especially in severe situations.

What is MPPT (Maximum Power Point Tracking)?

Maximum Power Point Tracking (MPPT) is a more advanced method for optimizing the power output of your solar panels. MPPT controllers continually monitor the voltage and current from your panels to determine the ideal power point, ensuring that your system generates the most power at all times. This is particularly useful in systems when the panel voltage is higher than the battery voltage.

Advantages of MPPT Charge Controllers

- **Higher Efficiency:** MPPT controllers may increase your solar system's efficiency by up to

30%, particularly in colder areas or when your panels are working under less-than-ideal conditions.

- **Flexibility:** MPPT controllers provide a wider choice of panel and battery combinations, making them ideal for both small and large systems.

- **Better Performance in Varying Conditions:** MPPT controllers perform in areas with variable sunshine or partial shadow, maximizing energy gathering.

Comparing Efficiency and Cost

When deciding between PWM and MPPT controllers, it is common to make a cost-efficiency trade-off. While PWM controllers are less expensive, MPPT controllers perform better, especially in bigger systems or under severe environmental conditions. For modest systems in sunny climes, a PWM controller may be sufficient, but for bigger installations or those in variable circumstances, an MPPT controller is typically the best investment.

Sizing and Selecting the Right Charge Controller

Selecting the appropriate charge controller for your system is important to ensuring that it runs effectively and securely. Here's how you go about it.

Determining System Voltage and Current Requirements

First, you must establish the voltage and current requirements for your solar system. This includes checking the voltage of your solar panels and battery bank, as well as the current that the panels will generate. The controller you select must be capable of handling the voltage and current of your system.

Matching Controller Capacity to Battery Bank

Your charge controller must be correctly sized for the capacity of your battery bank. Undersized controllers can result in inadequate charge, whilst big controllers may be superfluous and more expensive. The objective is to match the controller's amperage rating to the maximum output of your solar array.

Understanding Controller Ratings

Charge controllers exist in a variety of voltage, current, and power specifications. Make sure to choose a controller that not only meets your system's needs, but also has enough headroom to manage occasional surges or spikes in solar output. Consider the controller's temperature rating, particularly if your system will be exposed to harsh weather conditions.

Considerations for Different System Sizes

The size of your solar power system will also impact your selection of a charge controller. For tiny, basic systems, a PWM controller may be sufficient. However, as your system's complexity and power output increase, the advantages of using an MPPT controller become clearer. MPPT technology nearly always improves the efficiency of large systems, especially those with high voltage arrays.

Sizing and Selecting the Right Charge Controller

Understanding the purpose of a charge controller is the first step in selecting the appropriate one. At its core, a charge controller manages the voltage and current flowing from your solar panels to your battery bank, ensuring that the batteries are correctly charged and protected from overcharging. However, not all charge controllers are made equal, and picking the proper one involves careful evaluation of your system's individual needs.

- **Pulse Width Modulation (PWM) vs. Maximum Power Point Tracking (MPPT)**: These are the two main types of charge controllers. PWM controllers are simpler and more affordable, making them ideal for smaller systems with lower power needs. MPPT controllers, on the other hand, are more efficient and can handle higher voltages, making them suitable for larger, more complex systems. The choice between the two will depend on your budget, system size, and energy goals.

Determining System Voltage and Current Requirements

To correctly size a charge controller, you must first determine the voltage and current requirements of your system. This involves understanding the relationship between your solar panels, battery bank, and the devices you plan to power.

System Voltage: The voltage of your solar power system is typically 12V, 24V, or 48V. This is determined by the voltage of your battery bank and the solar panels' output. For example, if you have a 12V battery bank, your charge controller must be able to handle 12V. Larger systems often use 24V or 48V to reduce energy losses and wire sizes.

Current (Amperage) Requirements: The current your charge controller needs to handle is determined by the output of your solar panels and the overall energy demand of your system. You can calculate the required amperage by dividing the total wattage of your solar panels by the system voltage (e.g., a 1,200W system at 12V requires a charge controller that can handle 100A). This step is crucial to avoid overloading the controller, which could lead to system failure.

Matching Controller Capacity to Battery Bank

Once you've determined your system's voltage and current requirements, the next step is to match the charge controller's capacity to your battery bank. This ensures that the controller can effectively manage the energy flow between your solar panels and batteries.

Controller Capacity: Charge controllers are rated by the maximum current (in amps) they can handle. To ensure long-term reliability, select a controller with a capacity at least 25% higher than your calculated current requirements. For instance, if your system requires a 30A controller, consider choosing one rated for 40A to provide a margin of safety and allow for future expansion.

Battery Compatibility: Your charge controller must be compatible with the type of batteries in your system. Most controllers work with lead-acid, AGM, and lithium-ion batteries, but it's essential to check the manufacturer's specifications. The controller should also have settings that match the charging profile of your battery type, including absorption, float, and equalization stages.

Understanding Controller Ratings

Charge controllers come with a variety of ratings that you need to understand to make an informed decision.

Maximum Input Voltage: This is the highest voltage the controller can handle from your solar panels. Exceeding this voltage can damage the controller, so ensure your solar arrays voltage is within the controller's limits.

Maximum Output Current: This is the maximum current the controller can deliver to the battery bank. It should be greater than the calculated current requirements of your system to prevent overloading.

Efficiency: The efficiency rating of a charge controller indicates how much of the solar energy is effectively converted and used. MPPT controllers generally have higher efficiency ratings, often above 95%, compared to PWM controllers, which typically range from 70% to 80%.

Considerations for Different System Sizes

The size of your solar power system will heavily influence your choice of charge controller:

Small Systems: For smaller systems (e.g., less than 1 kW), a PWM controller may suffice. These systems often operate at 12V and have lower current requirements, making PWM a cost-effective choice.

Medium Systems: For medium-sized systems (1-3 kW), especially those using 24V or 48V battery banks, an MPPT controller is usually the best option. It can maximize energy harvest and handle higher voltages and currents.

Large Systems: Large systems (above 3 kW) almost always require an MPPT controller due to their efficiency and ability to handle the higher voltages and currents typical of these setups. Additionally, these systems may require multiple controllers working in parallel to manage the large power loads.

Programming and Configuring Charge Controllers

Configuring your charge controller correctly is crucial for the efficient operation of your solar system. While the specific programming steps can vary depending on the brand and model, the basic principles remain the same. Whether you're setting up a simple system or integrating multiple controllers, getting this right will ensure your batteries are charged optimally and your system remains stable.

Basic Programming Steps

Every charge controller needs to be programmed to align with your system's unique requirements. The process typically involves

Accessing the Controller's Menu: Start by navigating to the main menu of your charge controller. Most modern controllers feature an intuitive interface, often with an LCD display or mobile app for easy configuration.

Inputting System Specifications: Enter key details like the type and capacity of your solar panels, the voltage of your battery bank, and the total power output. This ensures the controller accurately manages the charge flow.

Setting Charge Parameters This is where the magic happens. Properly setting the charging parameters ensures your batteries are charged efficiently without being overworked.

Setting Charge Parameters

Charge parameters are critical because they dictate how the charge controller manages the flow of electricity to your batteries. Misconfigured settings can lead to undercharging or overcharging, both of which can shorten battery life.

Bulk Charge: This is the first stage where the controller delivers the maximum available current to charge the batteries quickly until they reach a specified voltage.

Absorption Charge: In this stage, the controller maintains a constant voltage to fully charge the batteries while tapering off the current. This prevents overcharging and minimizes stress on the batteries.

Float Charge: Once the batteries are fully charged, the controller reduces the voltage further to maintain the charge without causing damage. This is the maintenance phase, ensuring your batteries are always ready for use.

Configuring Battery Types and Settings

Different types of batteries—whether lead-acid, lithium-ion, or others—have distinct charging needs. Configuring your charge controller to match your battery type is essential for optimal performance.

1. **Selecting the Battery Type:** Your controller should have an option to choose the type of battery

in your system. This setting tailors the charge parameters to match the specific requirements of your battery bank.

2. **Setting the Battery Capacity:** Input the total amp-hour (Ah) capacity of your battery bank. This allows the controller to accurately manage the charge cycles.

3. **Temperature Compensation:** If your controller and batteries are equipped with temperature sensors, enable this feature to adjust the charging parameters based on temperature changes, preventing overcharging in hot conditions or undercharging in cold weather.

Understanding Display and Alerts

Most charge controllers have a display that shows real-time data regarding your system's performance. Familiarize yourself with these displays and alerts:

1. **Monitoring Charge Status:** The display will show the current charging stage—bulk, absorption, or float—as well as the voltage and current levels. Regularly check this to ensure everything is running smoothly.

2. **Understanding Alerts and Warnings:** Pay attention to any alerts or warnings displayed on the controller. These could indicate issues such as overvoltage, under voltage, or temperature problems. Addressing these promptly can prevent damage to your system.

3. **Utilizing Data Logs:** Some advanced controllers offer data logging features, allowing you to track the performance of your solar system over time. Reviewing this data can help you identify trends and optimize your setup.

Remote Monitoring and Control Features

In today's connected world, many charge controllers come with remote monitoring and control capabilities.

These features allow you to manage your solar system from anywhere, ensuring that you can keep an eye on your system's performance even when you're off-site.

1. **Connecting via Mobile Apps:** Many manufacturers offer mobile apps that sync with your charge controller. These apps provide real-time data and allow you to adjust settings remotely.

2. **Integrating with Smart Home Systems:** Some controllers can be integrated with broader smart home systems, enabling you to monitor and control your solar setup alongside other home automation features.

3. **Using Web Portals:** For more extensive systems, web portals offer a comprehensive overview of your solar setup. You can access detailed reports, adjust settings, and even troubleshoot issues through these platforms.

Integrating Multiple Charge Controllers in a System

In larger off-grid systems, a single charge controller might not be sufficient to handle the entire load. Integrating multiple charge controllers can be a viable solution, but it requires careful planning and configuration.

When to Use Multiple Controllers

Multiple charge controllers are typically needed when.

1. **High Power Requirements:** If your solar array's output exceeds the capacity of a single charge controller, adding more controllers can distribute the load.

2. **Multiple Battery Banks:** If your system includes several battery banks with different charge requirements, multiple controllers ensure each bank is managed optimally.

3. **System Expansion:** As your power needs grow, adding additional charge controllers can allow you

to expand your solar array without overloading your existing system.

Wiring and Configuration of Multiple Controllers

When integrating multiple controllers, proper wiring and configuration are crucial for balancing the load and ensuring efficient operation.

1. **Parallel vs. Series Configuration:** Decide whether to wire your controllers in parallel or series based on your system's design and power requirements. Parallel wiring is more common, as it allows each controller to operate independently.

2. **Connecting to a Common Battery Bank:** Ensure all controllers are properly connected to the same battery bank. Use bus bars or combiner boxes to manage the connections and maintain a balanced charge across the entire system.

3. **Synchronizing Charge Controllers:** If your system requires multiple controllers, they need to be synchronized to avoid conflicts. Some advanced

models offer synchronization features that automatically balance the load.

Monitoring and Troubleshooting Multi-Controller Systems

Managing multiple controllers can be challenging, but with the right approach, you can maintain an efficient and reliable system.

1. **Centralized Monitoring:** Use a centralized monitoring system to oversee the performance of each controller. This helps you spot any discrepancies in real-time and address them before they escalate.

2. **Troubleshooting Common Issues:** Familiarize yourself with common issues that arise in multi-controller systems, such as imbalance in power distribution or communication errors between controllers. Regular maintenance and prompt troubleshooting can prevent these issues from affecting your system's performance.

3. **Regular System Audits:** Conduct regular audits of your entire system, including each charge controller, to ensure everything is functioning as expected. Check for firmware updates and apply them to keep your controllers running smoothly.

Charge Controller Features and Enhancements

When selecting a solar charge controller, understanding its features and enhancements is a good point to maximizing the performance and longevity of your system.

Temperature Compensation

Temperature plays a significant role in how your batteries charge and discharge. Most charge controllers have temperature adjustment functions that automatically alter the charging voltage depending on the ambient temperature. This feature ensures that your batteries charge efficiently in various climates, protecting them from being overcharged in hot conditions or undercharged in cold weather.

Voltage Regulation

Voltage regulation is the primary function of a charge controller. It ensures that your batteries receive the correct amount of voltage, preventing overcharging or undercharging, which could damage the battery over time. Different charge controllers offer various levels of voltage regulation, from basic models that handle the essentials to advanced models that provide precise control and monitoring.

Load Control Features

Some charge controllers come with load control features, allowing you to connect and manage direct DC loads (like lights or fans) directly from the controller. This feature is particularly useful in off-grid setups where managing energy consumption is crucial. Load control features can help extend battery life by automatically disconnecting non-essential loads when the battery voltage drops below a certain threshold.

Battery Equalization

Battery equalization is a controlled overcharge that helps to equalize the charge across all cells in a battery bank. Some charge controllers have an automatic equalization feature that periodically performs this function, which can extend the life of your batteries by preventing sulfation a common cause of battery failure.

Charge Controller Safety and Maintenance

Proper safety and regular maintenance of your charge controller are essential for ensuring the longevity of your off-grid solar system.

Common Issues and Troubleshooting

Like any piece of technology, charge controllers can sometimes run into issues. Common problems include incorrect wiring, blown fuses, or improper settings. To troubleshoot, always start by checking the wiring connections, ensuring all cables are securely attached and in good condition. If the controller isn't functioning correctly, a factory reset can often resolve issues by restoring default settings.

Regular Maintenance Practices

Regular maintenance is a key to keeping your charge controller in top shape. This includes cleaning the terminals to prevent corrosion, checking for loose connections, and ensuring that the controller is operating within its specified temperature range. Keeping a log of maintenance activities can also help in diagnosing issues before they become serious problems.

Protecting the Charge Controller from Damage

To protect your charge controller from damage, make sure it's installed in a well-ventilated area, away from direct sunlight or exposure to the elements. Using surge protection devices can also help guard against voltage spikes that could damage the controller. It's wise to monitor the system for any signs of overheating or abnormal operation.

Warranty and Support Considerations

When choosing a charge controller, consider the warranty and support offered by the manufacturer. A solid warranty may provide you piece of mind and preserve your investment. It's also helpful to choose a manufacturer that offers robust customer support, as this can be invaluable when troubleshooting or needing replacement parts.

Understanding Charge Controller Specifications

When selecting a charge controller, it's important to understand the technical specifications to ensure compatibility with your solar system.

Voltage Ratings

The voltage rating of a charge controller indicates the maximum voltage it can handle from the solar panels. Ensure that the controller you choose matches the voltage of your solar panels and batteries. For instance, a 12V system requires a charge controller rated for 12V, though many controllers are capable of handling multiple voltage levels (e.g., 12V/24V/48V).

Current Ratings

The current rating, which is frequently given in amps, shows how much current the controller can manage from the solar panels to the batteries. To avoid overheating and failure, use a controller with a current rating that surpasses the peak current output of your solar array.

Efficiency and Losses

Efficiency is a measure of how much electricity the charge controller can convert from solar panels to batteries with no losses. Look for high-efficiency controllers to guarantee that as much solar energy as possible is stored in your batteries. Low-efficiency controllers can cause substantial power loss, lowering the overall efficacy of your system.

Certification and Standards Compliance

Finally, select a charge controller that fulfills the necessary certifications and requirements. Look for controllers that meet industry standards such as UL, CE, or TUV certification, which show that the device has been evaluated for safety and dependability. Compliance with these criteria may also be required for some warranties and refunds.

Future Trends in Charge Controllers

As the solar business expands, so does the technology that drives it. Charge controllers are getting increasingly complex, with a clear trend toward improved efficiency, integration, and usability. One of the most significant developments is the use of Maximum Power Point Tracking (MPPT) controllers. These gadgets are growing more complex, allowing for finer adjustments and better performance under varied environmental circumstances. As a result, MPPT controllers are projected to become the standard, even for smaller and less costly systems.

Another rising trend is the miniaturization of charge controllers. As technology advances, these gadgets become smaller and more compact, while maintaining power and efficiency. This makes them easy to integrate into varied solar configurations, whether it's for a little lodge or a huge off-grid residence.

Innovations and Technological Advances

The field of charge controllers is rife with innovation, driven by the need for more efficient, dependable, and user-friendly solutions. One intriguing innovation is the creation of adaptive charging algorithms. These clever algorithms tailor the charging process to the battery's state and usage patterns, prolonging battery life and increasing overall system efficiency.

Another technological advance is wireless connection, which allows consumers to remotely monitor and operate their solar power installations. Charge controllers may now be accessible using smartphones, tablets, and laptops that support Bluetooth and Wi-Fi. This innovation takes convenience to the next level, allowing real-time monitoring and changes from nearly anywhere.

Furthermore, multi-stage charging is becoming more advanced. This method gradually charges batteries, ensuring that they be charged efficiently without overloading or destroying them. The most recent charge controllers may adjust these steps based on the precise type of battery being used, whether it's lead-acid, lithium-ion, or another variety.

Emerging Features and Capabilities

The future of solar charge controllers is about not only enhancing what we have now, but also delivering whole new features that will change the way we utilize solar power. One such function is bidirectional charging, which enables the controller to charge and drain the battery. This capacity is especially beneficial in systems that employ vehicle-to-grid (V2G) technology or are linked to other renewable energy sources.

Another developing element is energy forecasting. Advanced charge controllers can forecast future energy generation potential based on meteorological data and prior performance. This helps consumers to better manage their energy use, especially in places with intermittent sunshine.

The incorporation of artificial intelligence (AI) is on the agenda. AI-powered charge controllers will be able to learn from user behavior and ambient circumstances, optimizing energy usage and storage automatically. This might dramatically improve the efficiency and dependability of off-grid systems, making solar energy more accessible and sustainable.

Integration with Smart Home Systems

Solar power systems evolve in tandem with the intelligence of dwellings. The integration of solar charge controllers with smart home systems is revolutionary, offering unrivaled control and efficiency. Smart home integration allows solar systems to work with other energy management equipment like thermostats, lighting, and appliances, ensuring that energy is used when and where it is most efficient.

For example, a smart solar system may prioritize charging during peak sunlight hours and distribute energy to different home appliances based on their usage patterns. If a smart thermostat detects that the home has reached a comfortable temperature, it may signal the charge controller to transfer power to other devices or back to the grid, thereby enhancing energy efficiency.

Voice control integration enables customers to run their solar power system with virtual assistants like Amazon Alexa or Google Home. Simple voice commands may initiate tasks such as monitoring battery levels, activating backup power, or even scheduling maintenance notices.

These advancements are not only handy; they represent a fundamental shift in how we interact with solar power systems. The integration of charge controllers with smart home technologies gets us closer to an energy management future that is easy, intuitive, and sustainable.

Solar charge controllers are the foundation of any off-grid system, balancing solar power production with battery storage. Understanding their characteristics, properly maintaining them, and selecting the appropriate specifications will ensure that your solar power system remains efficient, safe, and reliable for many years.

Are you ready to take responsibility of your solar system? Let's move on to the next phase of this book, where we will look at solar battery systems and how to improve their performance.

CHAPTER SIX

SOLAR BATTERY SYSTEMS

A dependable solar battery system is the foundation of any off-grid solar power system, providing energy storage to keep your lights and appliances operating even when the sun isn't shining. This chapter digs into the important features of solar battery systems; from knowing the types of batteries available to sizing and arranging your battery bank to meet your energy requirements.

Overview of Deep-Cycle Batteries

Deep-cycle batteries are designed to provide constant, long-term power, making them perfect for off-grid solar systems. Unlike vehicle batteries, which are designed for brief bursts of energy, deep-cycle batteries may be drained and recharged repeatedly without losing capacity. This capacity to sustain severe discharges makes them vital in solar power systems where energy demands vary between day and night.

Lead-Acid vs. Lithium-Ion Batteries

Choosing the proper battery type is important to the efficiency and lifespan of your solar system. The two basic possibilities are lead-acid and lithium-ion batteries, each having their own strengths and disadvantages.

Lead-Acid Mechanics

Lead-acid batteries are the oldest and most used form of rechargeable battery. They use a chemical interaction between lead plates and sulfuric acid to create energy. Solar power systems employ two types of lead-acid batteries: flooded lead-acid (FLA) and sealed lead-acid (SLA).

Types of Lead-Acid Batteries

1. **Flooded Lead-Acid (FLA) Batteries**: These are the most cost-effective but require regular maintenance, including checking electrolyte levels and ensuring proper ventilation to prevent hydrogen gas buildup.

2. **Sealed Lead-Acid (SLA) Batteries**: Also known as maintenance-free batteries, SLAs are sealed, so they don't require regular watering. They come in two main varieties: Gel and Absorbed Glass Mat (AGM). Gel batteries are ideal for deep discharges, while AGM batteries offer better performance at a higher cost.

Lithium-Ion Mechanics

Lithium-Ion (Li-Ion) batteries have surged in popularity due to their high energy density, longer lifespan, and minimal maintenance requirements. They work by moving lithium ions from the positive to the negative electrode during discharge and back during charging. Li-Ion batteries are more efficient than Lead-Acid, meaning they can store more energy for the same weight and volume.

Li-Ion vs. Lead-Acid

- **Cost**: Lead-Acid batteries are more affordable upfront, but Li-Ion batteries, though more expensive, often prove more cost-effective over time due to their longer lifespan and higher efficiency.

- **Efficiency**: Li-Ion batteries can reach up to 95% efficiency; compared to 70-85% for Lead-Acid, meaning more of the power you generate is stored for use.

- **Lifespan**: Li-Ion batteries typically last 10-15 years, compared to 5-7 years for Lead-Acid batteries.

- **Maintenance**: Lead-Acid batteries require regular maintenance, whereas Li-Ion batteries are virtually maintenance-free.

Understanding these differences helps you make an informed decision when selecting batteries for your system.

Sizing and Configuring Your Battery Bank

Once you've chosen the type of battery, the next step is to size and configure your battery bank to meet your specific energy needs.

Determining Battery Capacity

Battery capacity, measured in ampere-hours (Ah), influences the amount of energy that your batteries can store. To calculate the required capacity, you need to estimate your daily energy consumption in kilowatt-hours (kWh) and then convert that into ampere-hours based on your battery voltage.

Formula:

$$\text{Battery Capacity (Ah)} = \frac{\text{Daily Energy Consumption (kWh)}}{\text{Battery Voltage (V)}} \times \text{Days of Autonomy}$$

Configuring Batteries in Series and Parallel

Batteries can be connected in series, parallel, or a combination of both to achieve the desired voltage and capacity.

- **Series Connection**: Connecting batteries in series increases the voltage while keeping the capacity the same. For example, connecting two 12V batteries in series gives you 24V but with the same capacity as a single battery.

- **Parallel Connection**: Connecting batteries in parallel increases the capacity while keeping the voltage the same. For example, connecting two 100Ah batteries in parallel gives you 200Ah at the same voltage.

Choosing the right configuration is essential to match your inverter's input voltage and ensure your battery bank can meet your energy storage needs.

Battery Bank Sizing for Different Applications

Different applications require different battery bank sizes. For instance:

- **Small Off-Grid Cabins**: Might only need a few kWh of storage, so a small bank of Lead-Acid batteries might suffice.

- **Off-Grid Homes**: Will need a larger battery bank to provide enough energy for all household needs, often favoring Lithium-Ion batteries for their efficiency and longevity.

- **Emergency Backup Systems**: Requires a battery bank that can provide short-term power during outages, where quick discharge and recharge is the key to emergency backup systems.

Days of Autonomy and System Design

Days of autonomy refer to how many days your battery bank can power your home without any solar input. This is crucial for areas with frequent cloudy days or in winter when sunlight is limited.

Formula:

$$\text{Days of Autonomy} = \frac{\text{Battery Capacity (Ah)} \times \text{Battery Voltage (V)} \times \text{System Efficiency}}{\text{Daily Energy Consumption (kWh)}}$$

Designing your system with the appropriate days of autonomy ensures you have reliable power even during prolonged periods without sun.

Battery Monitoring and Maintenance

Proper monitoring and maintenance are essential for ensuring the longevity and performance of your solar batteries. Without consistent care, your batteries could degrade faster than expected, leading to reduced capacity and efficiency.

Battery Monitoring Tools

To keep an eye on your battery health, investing in the right monitoring tools is crucial. These tools allow you to track vital parameters such as voltage, current, state of charge (SoC), and temperature. Many modern monitoring systems provide real-time data and can alert you to potential issues before they become serious problems. Consider using:

- **Battery Monitors:** Devices that track energy flow and provide detailed insights into your battery's performance.

- **Smartphone Apps:** Many battery monitoring systems come with apps that allow you to check your battery status remotely.

- **Voltage Meters:** Simple yet effective tools to measure the voltage across your battery terminals, ensuring they remain within optimal levels.

Regular Maintenance Practices

Just like any other component in your solar power system, batteries require regular maintenance to function correctly. Here are some best practices:

- **Clean the Battery Terminals:** Dirt and corrosion on battery terminals can impede performance. Regularly cleaning them with a mixture of baking soda and water can prevent buildup.

- **Check Water Levels:** For lead-acid batteries, maintaining the correct water level is critical. Use distilled water and avoid overfilling.

- **Inspect for Damage:** Regularly check for any signs of damage, such as cracks or bulges in the battery casing. Addressing these difficulties as soon as possible will help to avoid worse problems.

- **Equalization Charging:** This process helps balance the charge across all cells in a battery, particularly in lead-acid types, preventing sulfation and extending battery life.

Troubleshooting Common Battery Issues

Even with diligent maintenance, issues can arise. Here are the basic procedures to troubleshoot some common difficulties:

- **Low Voltage:** If your battery voltage is consistently low, it could indicate a problem with your solar panels, charge controller, or battery capacity. Start by checking the input from your panels and the settings on your charge controller.

- **Overcharging:** Overcharging can cause batteries to overheat and degrade quickly. Ensure your charge controller is correctly set to avoid this issue.

- **Sulfation:** This occurs when lead-acid batteries are left undercharged for extended periods. Equalization charging can sometimes reverse mild sulfation.

Battery Health Improvement Strategies

Maintaining optimal battery health ensures that your system runs smoothly for years to come. Here are a few strategies:

- **Avoid Deep Discharge:** Regularly discharging your batteries too deeply can shorten their lifespan. Aim to keep your batteries above 50% charge whenever possible.

- **Temperature Control:** Batteries function optimally within a specified temperature range. Avoid exposing them to extreme heat or cold, and consider insulation or temperature-controlled environments if needed.

- **Regular Testing:** Perform periodic load tests to evaluate your battery's capacity and performance.

National Electric Codes for Battery Installation

Compliance with the National Electric Code (NEC) is essential for the safe and legal installation of your solar battery system. The NEC provides guidelines to ensure that all electrical installations are safe and standardized.

Battery Terminals and Connections

Proper connection of battery terminals is critical to prevent short circuits, corrosion, and potential fire hazards. The NEC recommends using corrosion-resistant terminals and ensuring all connections are tight and secure.

- **Use Proper Terminal Lugs:** These ensure a solid connection and reduce resistance.
- **Apply Anti-Corrosion Grease:** This helps prevent oxidation on the terminals.

Battery Compartment Design

The design of your battery compartment plays a crucial role in the safety and longevity of your battery system.

- **Ventilation:** Batteries, particularly lead-acid types, can emit hydrogen gas during charging. Ensure your battery compartment is well-ventilated to prevent gas buildup.

- **Accessibility:** Design your battery compartment so that batteries are easy to access for maintenance and replacement.

- **Thermal Management:** Consider the thermal management of your battery compartment, particularly in extreme climates. Insulation, ventilation, and cooling systems can help maintain optimal temperatures.

Grounding Requirements

Grounding your solar battery system is not just a recommendation—it's a requirement by the NEC to ensure safety. Proper grounding helps protect your system from lightning strikes and electrical faults.

- **Follow Local Codes:** Grounding requirements can vary depending on your location. Always consult local codes in addition to the NEC.

- **Use Ground Rods:** Drive ground rods into the earth and connect them to your system's grounding bus bar.

Compliance with NEC Standards

To ensure your system is up to code and safe, it's essential to follow all NEC standards related to battery installation:

- **Conduit Use:** NEC requires that certain battery cables be run through conduit to prevent damage.

- **Labeling:** Proper labeling of battery systems is required, especially for emergency responders.

- **Inspection:** Have your system inspected by a certified electrician to ensure it meets all NEC requirements.

Temperature Effects on Battery Performance

Temperature is one of the most significant factors influencing battery performance and lifespan. Both extreme heat and cold can impact how effectively your batteries function.

Impact of Temperature on Lead-Acid Batteries

Lead-acid batteries, commonly used in off-grid systems, are particularly sensitive to temperature changes. High temperatures can cause the electrolyte to evaporate more quickly, leading to reduced capacity and increased risk of battery damage. Conversely, cold temperatures can slow the chemical reactions inside the battery, diminishing its performance and reducing its ability to hold a charge.

To illustrate, imagine a hot summer day where the temperature in your battery storage area rises significantly. You may notice a decrease in battery efficiency and the requirement for more frequent charging. On the flip side, during a cold winter night, the battery might struggle to deliver power as efficiently, leading to potential energy shortages.

- **Managing Temperature in Battery Banks**

Proper management of battery temperature is essential for maintaining performance and extending lifespan. Ideally, batteries should be stored in a temperature-controlled environment where extremes are avoided. If this isn't possible, you can use insulation or shading to mitigate temperature fluctuations.

For instance, installing ventilation or cooling systems in battery storage rooms can help dissipate excess heat, while thermal blankets or heating pads can prevent batteries from getting too cold in winter.

Thermal Management Systems

Advanced solar systems often incorporate thermal management systems to regulate battery temperature automatically. These systems can include fans, heat exchangers, or climate-controlled enclosures that keep the battery bank at an optimal temperature range. Such systems are particularly beneficial for large setups or in areas with extreme temperatures.

Battery Lifecycle and Replacement

Batteries don't last forever, and knowing when to replace them is crucial for maintaining a reliable energy supply. Understanding battery lifespan and recognizing signs of deterioration will help you make informed decisions about replacements.

Estimating Battery Lifespan

The lifespan of a battery is influenced by various factors, including temperature, usage patterns, and maintenance. Lead-acid batteries typically last between 3 to 5 years, while lithium-ion batteries can last up to 10 years or more. Regular monitoring and proper care can maximize the lifespan of your batteries.

Consider a case where you've been using your battery bank for several years. Regularly checking the battery's state of charge and performance will help you gauge whether it's nearing the end of its useful life.

- **Indicators for Battery Replacement**

Signs that your batteries may need replacing include a noticeable drop in performance, frequent charging, or difficulty holding a charge. Additionally, physical indicators such as swelling, corrosion, or leaks can signal that the batteries are deteriorating.

For example, if you find that your solar system can no longer meet your daily energy needs despite being fully charged, it might be time to assess the condition of your batteries and consider replacement.

Recycling and Disposal of Old Batteries

It's essential to properly dispose of and recycle old batteries to protect environmental sustainability. Many battery types, particularly lead-acid batteries, contain hazardous materials that require careful handling.

Most communities have designated recycling centers for batteries. Contact local waste management services or battery retailers to find out the best way to dispose of your old batteries. Some manufacturers provide recycling and take-back programs for their products.

Battery Safety Protocols and Precautions during Installation

Proper installation is the key to both the performance and safety of your solar battery system. Here are some critical safety precautions to follow:

1. **Read the Manufacturer's Instructions:** Each battery type comes with its own set of installation guidelines. Adhering to these instructions ensures compatibility and safety.

2. **Use Proper Tools and Gear:** Wear appropriate safety gear, such as gloves and safety glasses, and use the right tools to prevent accidents.

3. **Ensure Proper Ventilation:** Many batteries, particularly lead-acid types, emit gases that can be hazardous. Install your batteries in a well-ventilated area to prevent the buildup of harmful gases.

4. **Verify Electrical Connections:** Double-check all connections for correct polarity and tightness to prevent short circuits and potential fires.

Handling and Storage Best Practices

Handling and storing batteries correctly is also important to their longevity and your safety:

1. **Store Batteries in a Cool, Dry Place:** Excessive heat and humidity can degrade battery performance. Keep them in a steady atmosphere to extend their lives.

2. **Avoid Physical Damage:** Batteries should be handled with care to avoid any physical damage, which could lead to leaks or other issues.

3. **Keep Batteries Clean:** Ensure that battery terminals are clean and free from corrosion to maintain a good connection.

Emergency Procedures and Safety Measures

In case of emergencies, knowing what to do can prevent accidents and injuries.

1. **Know the Location of the Battery Disconnect Switch:** Ensure you can quickly access and operate the disconnect switch to cut off power in an emergency.

2. **Have a Spill Kit Ready:** For lead-acid batteries, have a spill kit available to handle any accidental leaks safely.

3. **Understand Battery Symptoms:** Recognize signs of battery issues, such as swelling, excessive heat, or strange odors, and address them immediately.

Innovations and Emerging Technologies

Battery technology is evolving rapidly, offering improved performance, efficiency, and safety. We are to identify some of the newest developments.

1. **Lithium-Ion Batteries:** These batteries are becoming the go-to choice for off-grid solar

systems due to their high energy density, longer lifespan, and lower maintenance requirements compared to traditional lead-acid batteries.

2. **Solid-State Batteries:** Still in the development stage, solid-state batteries promise higher safety and energy density by replacing liquid electrolytes with solid materials.

3. **Flow Batteries:** Offering long cycle life and scalability, flow batteries are being explored for larger off-grid systems and grid-scale applications.

Future Trends in Battery Systems

Looking ahead, several trends are shaping the future of solar battery systems:

1. **Increased Energy Density:** Future batteries will continue to offer higher energy density, allowing for more compact and powerful storage solutions.

2. **Enhanced Recycling:** Innovations in recycling processes will improve the sustainability of battery systems by reducing waste and reusing materials.

3. **Smart Battery Management:** Advanced battery management systems will provide real-time data and analytics, optimizing performance and extending battery life.

Integrating New Technologies into Existing Systems

As new technologies emerge, integrating them into your existing solar battery system can enhance performance:

1. **Upgrade Compatibility:** Ensure that any new technology is compatible with your current system components before integration.

2. **Professional Installation:** For complex upgrades, consider consulting with a professional to ensure proper integration and avoid potential issues.

3. **Monitor System Performance:** After integrating new technologies, regularly monitor your system

to ensure it operates efficiently and addresses any issues promptly.

In this chapter, here we'll explore the fundamentals of off-grid inverters, including their types, technologies, and how to choose the right one for your system.

Next

CHAPTER SEVEN

OFF-GRID INVERTERS

Inverters are the core of your solar system, turning direct current (DC) from your solar panels into alternating current (AC) to power your home appliances. This chapter will walk you through the fundamentals of off-grid inverters, explaining their many types, technologies, and how to select the best one for your system.

Types of Inverters Grid-Tied, Hybrid, and Off-Grid

1. Grid-Tied Inverters

Grid-tied inverters are intended for systems that are linked to the electrical grid. They communicate with the grid to return extra energy. While ideal for grid-connected systems, they are unsuitable for off-grid setups since they require a grid connection to work.

2. Hybrid Inverters

Hybrid inverters provide versatility by operating as both grid-tied and off-grid inverters. They can store surplus energy in batteries while also allowing you to draw from the grid as needed. Hybrid inverters are appropriate for systems that may move from grid-connected to off-grid modes.

3. Off-Grid Inverters

Off-grid inverters are built particularly for systems that are not linked to the grid. They work independently, converting DC electricity from your solar panels or battery bank to AC power. These inverters offer a steady power supply, even in distant places.

Inverter Technologies: Modified vs. Pure Sine Wave

1. Modified Sine Wave Inverters

Modified sine wave inverters are a low-cost solution that can power a wide range of basic appliances. They generate a stepped waveform that resembles a sine wave but is not as smooth. While they function well for simple devices, sensitive electronics may face challenges like as buzzing or diminished performance.

2. Pure Sine Wave Inverters

Pure sine wave inverters generate a smooth and continuous waveform that is similar to the electricity provided by the grid. This kind is perfect for delicate devices and appliances that demand a consistent and clean power supply. Despite being more costly, they offer superior performance and safeguard your equipment.

Sizing and Selecting the Right Inverter

Choosing the appropriate inverter is always important to maximizing your solar power system. Begin by determining the total wattage of the devices you intend to operate simultaneously. Ensure that the inverter's continuous power rating surpasses this limit. Consider the peak power rating, which should allow any brief periods of significant power usage, such as running a refrigerator.

Understanding Inverter Power Ratings and Efficiency

1. Power Ratings

Inverters are graded according to their continuous and peak power outputs. The continuous power rating indicates the maximum load that the inverter can handle constantly. Peak power ratings show brief bursts of increased power, which are critical for appliances that demand more energy during starting.

2. Efficiency

Inverter efficiency refers to how well an inverter converts DC electricity to AC power. Look for inverters with excellent efficiency ratings, often greater than 90%. Higher efficiency implies that less energy is lost during conversion, resulting in more power for your usage.

Off-Grid Inverter System Integration

Integrating an off-grid inverter into your system requires series of key steps:

1. **Connect to Battery Bank**: Connect the inverter to your battery bank, ensuring adequate polarity and tight connection.

2. **Connect to Solar Panels**: Connect the solar panels to the DC input terminals of your inverter if it has a built-in charge controller.

3. **Connect to Load**: Connect the AC output of the inverter to your electrical panel or directly to the equipment you want to power.

Inverter Features and Specifications

1. DC Input Voltage Range

Check that the inverter's DC input voltage range fits your solar panel and battery arrangements. A mismatch might result in inefficient functioning or harm.

2. AC Output Voltage and Frequency

Inverters often produce alternating current in conventional voltages (e.g., 120V or 240V) and frequencies (e.g., 50Hz or 60Hz). Ensure that the inverter's output meets the requirements of your equipment.

3. Standby and Idle Power Consumption

Consider the inverter's standby or idle power consumption, which is the amount of power used while it is not actively converting electricity. Lower standby usage improves overall system efficiency.

4. Thermal Management and Cooling

Inverters create heat while operating. Look for models with adequate cooling systems, such as fans or heat sinks, to avoid overheating and provide consistent performance.

Inverter Installation and Setup

1. Mounting and Positioning

For best performance and endurance, your inverter must be properly mounted. Select a place that is:

- **Well-Ventilated**: Inverters produce heat, thus they should be positioned in a cool, dry area with good ventilation to avoid overheating.

- **Accessible**: Make sure you have easy access to the inverter for maintenance and monitoring.

- **Protected**: Avoid sites prone to harsh weather conditions or intense sunshine, as they might influence the inverter's effectiveness.

Mounting Tips:

- Use the inverter manufacturer's supplied brackets or mounting hardware.

- Attach the inverter to a firm surface to prevent vibrations and movement.

- Ensure the inverter is level to prevent internal damage and ensure proper cooling.

2. Wiring and Connections

Proper wiring is important for safety and efficiency. Follow these steps for correct wiring:

- **Connect to Batteries**: Ensure that the inverter's positive and negative terminals are connected to the appropriate batteries. Use the recommended wire gauge to accommodate the current load.

- **Connect to Solar Panels**: If your inverter accepts direct input from solar panels, connect them according to the manufacturer's specifications. Use proper connectors, and make sure all connections are secure.

- **Connect to Load**: Connect the inverter's AC output terminals to the domestic wiring or load. To avoid overloading, ensure that the load falls within the inverter's capability.

Wiring Tips:

- Use high-quality, properly rated cables and connections.

- • Maintain tidy and ordered wiring to prevent damage or interference.

- • Check all connections for security and accurate configuration before powering on the machine.

3. Safety Precautions

When installing and configuring your inverter, safety is the most important consideration. Follow these precautions:

- **Turn off Power**: To avoid electrical shocks, switch off the power before working on any wiring or connections.

- **Use Personal Protective Equipment (PPE)**: Wear insulated gloves and safety eyewear to protect you from electrical risks.

- **Follow Manufacturer Instructions**: To prevent warranty voiding and for safe operation, properly follow the manufacturer's installation and setup directions.

- **Grounding**: Ensure that the inverter is correctly grounded to avoid electrical failures and increase safety.

Monitoring and Maintenance

1. Inverter Monitoring Systems

Modern inverters frequently come bundled with monitoring devices that enable you to track performance and address concerns.

- **Display Panels**: Many inverters include built-in displays that give real-time information on system performance, such as voltage, current, and power output.

- **Remote Monitoring**: Some inverters provide remote monitoring via applications or online interfaces, allowing you to check system status and get alerts from anywhere.

Monitoring Tips:

- Regularly examine your inverter's display panel or remote monitoring system for proper operation.

- Respond to any warning lights or error messages swiftly.

2. Routine Maintenance Tasks

Routine maintenance is vital for keeping your inverter functioning properly:

- **Clean the Inverter**: Dust and dirt may collect on the inverter. Regularly wipe down the outside with a dry cloth to maintain it clean and free of obstacles.

- **Check Connections**: Inspect all wiring and connections at regular intervals for evidence of wear, corrosion, or loose fittings.

- **Inspect Ventilation**: Ensure that ventilation slots are free of impediments to promote optimum airflow and cooling.

Maintenance Tips:

- Conduct regular maintenance inspections, ideally every six months, to maintain your inverter in good condition.

- Consult the manufacturer's maintenance schedule and guidelines for precise suggestions.

3. Troubleshooting Common Issues

Even with appropriate installation and maintenance, you may experience problems with your inverter. These are some common challenges and solutions:

- **Inverter not powering On**: Check the connections to ensure they are secure, and make sure the batteries are properly charged. Check for tripped breakers and blown fuses.

- **Low Power Output**: Make sure the solar panels are producing enough electricity and that there are no obstacles or shading difficulties. Also, look for any defects or errors reported by the inverter's display or monitoring system.

- **Overheating**: Check that the inverter is in a well-ventilated location and that no ventilation holes are blocked. Clean the inverter and avoid exposing it to extreme heat sources.

Troubleshooting Tips:

- Refer to the inverter's user manual for troubleshooting tips relevant to your model.

- Consider seeking expert help if difficulties continue or get complex.

Battery Compatibility and Management

Matching Inverters with Battery Types

Battery Charging and Discharging Considerations

System Efficiency and Performance

Off-grid inverters are vital for increasing the efficiency of your solar power system. It transforms the direct current (DC) generated by your solar panels into alternating current (AC) for your appliances. The efficiency of your inverter has a direct impact on how much electricity is available for usage from your solar panels.

A high-quality inverter should have high conversion efficiency, usually 90% or more. This implies that just a fraction of the energy is wasted throughout the conversion process. To ensure that your inverter is performing optimally, check its efficiency ratings on a regular basis and compare them to the manufacturer's standards. Keep in mind that even the greatest inverters may wear down over time, so monitoring performance is important.

Optimizing Inverter Performance

To get the most out of your inverter, be sure it's performing optimally. Begin by verifying that it is appropriately sized for your system. An inverter that is too tiny will not meet your energy requirements, but one that is too large may waste electricity and be less efficient. Match the inverter capacity to the combined output of your solar panels and your household energy use.

Also, keep your inverter clean and well-ventilated. Dust and dirt can hinder its operation, resulting in overheating and decreased efficiency. To ensure ideal working conditions, keep your inverter in a cool, dry environment with enough circulation.

Maximizing Energy Harvesting

Maximizing energy harvesting is ensuring that your solar panels and inverter work together efficiently to catch and utilize as much solar energy as feasible. This may be accomplished by ensuring that your panels are properly positioned to face the sun and that any shade is minimized.

Inverters that provide maximum power point tracking (MPPT) are very beneficial. MPPT technology maximizes the amount of energy harvested from your solar panels by altering the electrical operating point to meet the highest power output. This means more energy captured and more power for you to use.

Advanced Inverter Features

Modern off-grid inverters have a number of innovative features aimed to improve your solar power system. These can include:

- **Battery Management**: Many inverters come with integrated battery management systems that monitor battery health, charge levels, and performance.

- **Dual-Input Capability**: Some inverters can accept inputs from both solar panels and wind turbines, making them adaptable for multi-source systems.

- **Power Backup Functions**: Advanced inverters can smoothly transition to backup power sources if the solar energy is inadequate.

Familiarize yourself with these aspects and pick an inverter that suits the unique demands and aims of your solar system.

Smart Grid Capabilities

Inverter technology improves alongside the smart grid. Smart grid-capable inverters can interface with the grid, optimizing power distribution and increasing overall system efficiency. They may change their operations in real time, adapt to grid circumstances, and even return excess energy to the grid when necessary. This link guarantees that your solar power system can integrate smoothly with the larger energy network, delivering dependability and efficiency.

Remote Monitoring and Control

One of the most significant advances in inverter technology is the ability to monitor and manage your system remotely. Many contemporary inverters have applications or internet platforms that enable you to monitor system performance, diagnose problems, and make changes from anywhere.

Remote monitoring gives piece of mind since it allows you to examine the health of your system and respond to concerns without having to be physically there. This functionality is especially beneficial for off-grid situations where access to the system may be limited.

Integration with Renewable Energy Sources

Integrating different renewable energy sources can improve the resilience and efficiency of an off-grid system. Many inverters are intended to work with sources other than solar panels, such as wind turbines and hydroelectric systems.

Combining several renewable energy sources ensures a more constant power supply and reduces dependency on a single source. This integration requires an inverter capable of accepting inputs from several types of generators; thus, pick an inverter that supports your unique arrangement.

Case Studies and Examples

1. The Johnson Family's Remote Cabin

Background: The Johnsons reside in a lonely cabin in the Appalachian Mountains, far away from the nearest electricity grid. They desired a sustainable solution for their energy needs, stressing independence and minimizing the environmental effect.

Setup: The Johnsons opted for a 3,000-watt pure sine wave inverter for their off-grid installation. This sort of inverter was chosen to guarantee that their sensitive electrical gadgets, such as their laptop and TV, ran smoothly and without interference or harm.

Outcome: The Johnsons were able to enjoy consistent electricity for their cabin's lights, refrigerator, and minor appliances thanks to the pure sine wave inverter. They achieved self-sufficiency and minimized their need on backup generators by combining it with a powerful battery bank and solar panels.

Lesson: For off-grid systems with sensitive electronics, investing in a pure sine wave inverter is critical for consistent and clean power supply.

2. GreenTech Community Center

Background: GreenTech, a community center in rural New Mexico, wanted to develop a green energy model for teaching reasons. They need a system that could manage both high and low power demands, including lighting, computers, and a small workshop.

Setup: They erected a 5,000-watt hybrid inverter, which was combined with a solar array and backup generator. The hybrid inverter was chosen for its ability to handle both solar electricity and generator input.

Outcome: The hybrid inverter efficiently controlled electricity from both sources, ensuring uninterrupted operation even on overcast days. The community center is currently a working example of renewable energy and sustainability.

Lesson: Hybrid inverters are perfect for systems that require a variety of energy sources, offering a dependable and adaptable answer to changing energy demands.

3. The Thompson Family's Tiny House

Background: The Thompsons planned to turn a tiny house into an off-grid residence and required an inverter that would suit in their small area and budget.

Setup: They chose a 1,000-watt modified sine wave inverter to meet their low power requirements and limited budget. This inverter powered the essential appliances and illumination.

Outcome: The customized sine wave inverter supplied enough electricity for the Thompsons' tiny abode, allowing them to live peacefully off-grid. However, they saw occasional problems with specific electrical gadgets, which encouraged them to contemplate upgrading to a pure sine wave inverter in the future.

Lesson: A modified sine wave inverter can be a cost-effective option for small-scale off-grid systems with minimal power demand, but possible restrictions must be considered.

Typical Off-Grid Inverter Setups

1. Basic Solar-Powered System

Components:

- **Inverter Type:** Modified Sine Wave
- **Power Rating:** 1,000 to 3,000 watts
- **Battery Bank:** 200 to 400 amp-hours
- **Solar Panels:** 200 to 400 watts

Description: This setup is ideal for small off-grid homes or cabins with minimal power needs. The modified sine wave inverter is cost-effective and sufficient for basic appliances and lighting.

2. Hybrid System

Components:

- **Inverter Type:** Hybrid
- **Power Rating:** 3,000 to 5,000 watts
- **Battery Bank:** 400 to 800 amp-hours
- **Solar Panels:** 500 to 1,000 watts
- **Generator:** Optional backup

Description: A hybrid system combines solar power with a generator or grid backup. It's suitable for medium-sized homes or community centers requiring a more reliable power source and the flexibility to manage multiple inputs.

3. High-Capacity System

Components:

- **Inverter Type:** Pure Sine Wave
- **Power Rating:** 5,000 to 10,000 watts

- **Battery Bank:** 800 to 1,200 amp-hours

- **Solar Panels:** 1,000 to 2,000 watts

- **Generator:** Optional backup

Description: This setup is designed for larger off-grid homes or small businesses with high power demands. The pure sine wave inverter ensures clean power for sensitive electronics and appliances, and the large battery bank supports extended autonomy.

Real-World Applications and Success Stories

1. Eco-Lodge in Costa Rica

Application: An eco-lodge in Costa Rica utilized a 7,000-watt pure sine wave inverter in combination with a large solar array and battery bank. The goal was to provide a sustainable energy source for guests in a remote rainforest location.

Success Story: The lodge successfully operated all amenities, including air conditioning, lighting, and small kitchen appliances, without relying on a diesel generator. The system proved reliable and significantly reduced the lodge's carbon footprint.

Takeaway: Large-scale solar power systems can support diverse energy needs in remote locations, enhancing sustainability and reducing environmental impact.

2. Urban Rooftop Farm

Application: A rooftop farm in a bustling city installed a hybrid inverter system to manage power from solar panels and a small wind turbine. The setup aimed to support urban agriculture and provide a green energy solution.

Success Story: The hybrid system efficiently managed power from both solar and wind sources, powering irrigation systems, grow lights, and climate controls. The rooftop farm became a model for urban sustainability and energy efficiency.

Takeaway: Hybrid inverters are versatile solutions for integrating multiple renewable energy sources, ideal for urban settings where space and resources are limited.

CHAPTER EIGHT

SOLAR WIRING & OVERCURRENT PROTECTION

When it comes to putting up an off-grid solar power system, correct wiring and overcurrent protection are very important. Here in this chapter we' will walk you through the process of choosing wire materials, explaining various forms of insulation, and guaranteeing your system's safety and efficiency. Let's look at the key parts of solar wiring and overcurrent safety.

Selecting the Right Wiring Materials

Choosing the appropriate wire materials is critical to the safety and functionality of your solar power system. The wire you choose must manage the electrical load securely and resist environmental conditions.

Types of Conductor Materials

Conductor materials are the core of your wiring system. Here are the common types used in solar installations:

- **Copper:** Copper, known for its high conductivity and flexibility, is a common material for solar wiring. It provides low resistance and is very dependable for both AC and DC circuits.

- **Aluminum:** Though less conductive than copper, aluminum is a lighter and less expensive alternative. It is frequently used in bigger installations, but requires specific handling to assure secure connections.

Insulation Types and Ratings

Insulation shields the conductor from environmental hazards and electrical problems. Select insulation that meets your system's voltage and temperature requirements:

- **THHN (Thermoplastic High Heat-resistant Nylon-Coated):** This insulation is ideal for dry regions and provides high protection against heat and chemicals.

- **THWN-2 (Thermoplastic Heat and Water-resistant Nylon-coated):** Suitable for wet conditions, THWN-2 provides extra protection against water and heat.

- **UV-Resistant Insulation:** For outdoor applications, UV-resistant insulation prevents degradation from sunlight exposure.

Conductor Color Codes

Correctly color-coded cables facilitate identification and assure safety during installation. This is a broad guide:

- **Black:** Typically used for negative or ungrounded conductors.

- **Red:** Often used for positive or live conductors.

- **White or Gray:** Commonly used for neutral conductors.

- **Green or Bare:** Reserved for grounding conductors.

Always check local codes and standards, as color codes can vary.

Wire Sizing Principles and Voltage Drop Calculations

Proper wire sizing is essential to maintain efficiency and safety. Incorrectly sized wires can lead to overheating, energy loss, and potential system failure.

Principles of Wire Sizing

Wire sizing is based on several factors including:

- **Current Rating (Ampacity):** Choose wires that can handle the maximum current expected in your system. Ampacity is determined by the wire's gauge, material, and insulation type.

- **Ambient Temperature:** Higher temperatures can reduce a wire's ampacity. Adjust wire size accordingly to prevent overheating.

Ampacity and Ambient Temperature Adjustment

Voltage drop occurs when electrical resistance causes a loss of voltage along the wire. Excessive voltage drop can reduce the efficiency of your solar system and lead to performance issues.

- **Calculating Voltage Drop:** Use the formula Vd=I×RV_d = I \times RVd=I×R, where VdV_dVd is voltage drop, III is current, and RRR is resistance. Keep voltage drop within acceptable limits (typically less than 3% of the system voltage) to ensure efficiency.

Voltage Drop and Its Impact on System Efficiency

When multiple conductors are bundled in a conduit, they generate heat, which can affect their ampacity. Follow these guidelines:

- **Derating Factors:** Apply derating factors as specified in the National Electrical Code (NEC) to account for heat buildup. This often involves reducing the ampacity of each conductor based on the number of conductors in the conduit.

- **Spacing and Ventilation:** Ensure proper spacing and ventilation within conduits to help dissipate heat and maintain safe operating temperatures.

Adjusting Ampacity for Multiple Conductors in a Conduit

When multiple wires are grouped together in a conduit, they generate heat, which can affect their ampacity.

1. **De-rating Factors:** Use the National Electrical Code (NEC) de-rating factors to adjust the ampacity of wires in a conduit. This involves reducing the ampacity of each wire based on the number of conductors and the type of conduit.

2. **Provide Adequate Ventilation:** Ensure that the conduit is not overly packed, which can impede heat dissipation. Proper ventilation and spacing help maintain wire integrity and performance.

Overcurrent Protection Devices (OCPD) and Their Sizing

Overcurrent Protection Devices are crucial in protecting your solar power system from excessive currents that can cause overheating, damage components, or even start fires. Understanding how to choose and size these devices appropriately is essential for a safe and effective solar setup.

Types of Overcurrent Protection Devices

1. **Fuses:** These are designed to blow or break the circuit when the current exceeds a certain level. They are a simple and reliable form of overcurrent protection.

2. **Circuit Breakers:** These automatically switch off when they detect an overcurrent, and they can be reset after tripping. Circuit breakers offer convenience and can protect against both overloads and short circuits.

3. **Integrated Protection Devices:** Some solar equipment comes with built-in overcurrent

protection, which combines functionality and convenience into a single unit.

Sizing OCPDs for Different System Components

- **String Fuses:** Protect individual strings of solar panels. They should be rated to handle the maximum current that could be produced by the string, considering the worst-case scenario.

- **Combiners to the Charge Controller:** Fuse or breaker protection is needed here to safeguard the connection between the solar panel combiner box and the charge controller.

- **Charge Controller to the Busbar (Battery):** Size the protection device to handle the maximum current output of the charge controller and the potential short-circuit current.

- **Battery to Busbar (Inverter):** Protect this connection to prevent damage to the inverter and battery due to excessive currents.

- **Inverter's AC Output Protection:** Ensure that the inverter's output is protected to prevent overcurrent conditions that could affect household appliances or grid-tie systems.

String Fuses and Their Placement

String fuses are an essential part of solar panel protection. They should be installed in the positive lead of each string of solar panels to protect against the possibility of a short circuit or overcurrent within the panel strings. Proper placement ensures that if a fault occurs, only the affected string is isolated, maintaining the operation of the remaining system.

Combiners to the Charge Controller

The **combiner box** aggregates the outputs from multiple strings of solar panels. Proper overcurrent protection here is vital to prevent faults from propagating to the charge controller. Use fuses or circuit breakers rated for the total current that the combiner will handle.

Charge Controller to the Bus bar (Battery)

When connecting the charge controller to the battery bank through a busbar, overcurrent protection helps prevent damage to both the controller and batteries. Select a fuse or breaker that matches the maximum output of the charge controller and provides a margin for safety.

Battery to Bus bar (Inverter)

Battery connections to the inverter require robust overcurrent protection to shield the inverter and battery bank from surges or faults. Ensure that the protection device is appropriately rated for the battery bank's maximum current output and the inverter's input requirements.

Inverter's AC Output Protection

Protecting the AC output of the inverter is important for preventing harm to household appliances and other connected equipment. Use circuit breakers or fuses designed for the inverter's output capacity to safeguard against overcurrent conditions.

Equipment Grounding and GFCI

Grounding is a fundamental safety measure in solar power systems. Proper grounding helps to prevent electrical shock hazards, ensures system stability, and protects equipment from lightning strikes or surges.

Importance of Grounding in Solar Systems

Grounding provides a safe path for electrical faults and prevents the build-up of dangerous voltages. It helps ensure that any fault currents are safely directed away from users and equipment, reducing the risk of electrical fires or shocks.

Types of Grounding Methods

1. **System Grounding:** Connects the system's neutral point to the earth to provide a reference point for the system voltage.

2. **Equipment Grounding:** Involves grounding all metal parts of the equipment to prevent electrical shock and ensure safety.

3. **Grounding Electrode System:** Uses grounding rods or plates to establish a low-resistance path to the earth.

Ground Fault Circuit Interrupter (GFCI) Usage

GFCIs are designed to detect ground faults and disconnect power to prevent electrical shock. They are crucial for protecting both AC and DC systems, particularly in areas prone to moisture and where accidental contact with conductive surfaces may occur.

Grounding for AC and DC Systems

- **AC Systems:** Ensure proper grounding of the inverter's AC output and all related circuits. This protects users from electric shock and ensures system stability.

- **DC Systems:** Ground the positive and negative leads where required, following system design and safety codes. Ensure that all DC components, including the panels, charge controllers, and batteries, are properly grounded.

Proper Wire Connections and Terminations

The foundation of a reliable solar power system begins with secure and correct wire connections. These connections must be robust enough to handle the electrical load while minimizing the risk of electrical faults.

Types of Wire Connectors and Terminals

Choosing the right connectors and terminals is crucial. Here are some common types:

- **Crimp Connectors:** Ideal for creating solid, reliable connections. Crimp connectors come in various sizes and types, such as ring terminals and spade terminals, each suited for specific applications.

- **MC4 Connectors**: These are standard in solar installations. They provide a waterproof and secure connection between solar panels and other system components.

- **Bus bars**: Used to connect multiple wires in a compact, organized manner. Bus bars help distribute power efficiently across your system.

- **Junction Boxes**: These enclosures house electrical connections and protect them from environmental elements.

Techniques for Secure Wire Connections

Achieving a reliable connection involves more than just choosing the right connectors:

- **Proper Crimping**: Use a crimping tool to securely attach connectors to wires. Ensure that the crimp is tight and that no strands of wire are exposed.

- **Soldering**: For some connections, soldering can provide an even more reliable bond. Ensure you use the correct soldering technique and materials to avoid weak connections.

- **Heat Shrink Tubing**: After connecting wires, use heat shrink tubing to insulate and protect the connections. This helps prevent shorts and keeps connections secure.

Preventing Corrosion and Ensuring Longevity

Corrosion can significantly impact the performance and safety of your wiring. Here's how to prevent it:

- **Use Corrosion-Resistant Materials**: Opt for connectors and terminals made from materials resistant to corrosion, such as stainless steel or tinned copper.

- **Seal Connections**: Apply dielectric grease to connections to repel moisture and prevent corrosion. Ensure all connections are properly sealed with waterproof enclosures or heat shrink tubing.

Testing and Verifying Connections

Regular testing is essential to ensure your wiring is functioning correctly:

- **Visual inspection**: Check for evidence of wear, damage, or loose connections on a regular basis. Make that all connections are secure and free of corrosion.

- **Electrical Testing**: Use a multi meter to test for continuity and check voltage levels. This helps identify any issues with the connections or overall system performance.

Circuit Protection and Safety Measures

Circuit protection is essential to keep your system safe from overloads and short circuits. Implementing the appropriate safety precautions can improve the dependability and lifetime of your solar power system.

Understanding Circuit Protection Needs

Different parts of your system require specific protection measures:

- **Fuses**: Fuses protect circuits by breaking the connection if the current exceeds a certain level. They are typically used in series with the load.

- **Circuit Breakers**: These automatically disconnect the circuit when an overload or short circuit is detected. They are reusable and can be reset once the issue is resolved.

- **Overcurrent Protection Devices (OCPD)**: These devices prevent excessive current from damaging your wiring and equipment. They come in various forms, including fuses and breakers.

Implementing Safety Measures in Wiring

Working with electrical systems requires a high level of safety:

- **Proper Sizing**: Ensure that wires are sized correctly for the current they will carry. Undersized wires can overheat and cause fires.

- **Secure Mounting**: Use appropriate cable management systems to prevent wires from moving, rubbing against surfaces, or becoming damaged.

- **Grounding**: Properly ground your system to prevent electrical shocks and ensure safe operation.

Regular Inspection and Maintenance Practices

Maintaining your wiring system is crucial for long-term reliability:

- **Routine Checks**: Regularly inspect all wiring and connections for signs of wear, damage, or corrosion.

- **Cleaning**: Keep wiring and connectors clean from dust, dirt, and other debris that could impact performance.

- **Documentation**: Maintain records of your system's wiring layout and any maintenance performed. This can be extremely useful for troubleshooting and future improvements.

Integration with Solar Components

When setting up your off-grid solar power system, proper wiring is crucial for ensuring that all components work harmoniously and efficiently. This chapter will guide you through the essential steps to connect your solar panels, charge controllers, batteries, and inverters. You'll also learn how to protect your system from overcurrent issues, which is vital for both safety and performance.

Connecting Panels to the Combiner Box

The combiner box is a central hub where the wires from your solar panels converge before they head to charge controller. To connect your panels to the combiner box:

1. **Preparation**: Make sure all connections are made with the system powered off. Use appropriate tools and safety gear.

2. **Wiring**: Connect the positive and negative wires from each solar panel to the respective terminals in the combiner box. Ensure that you follow the manufacturer's guidelines for wire sizes and connections.

3. **Safety Check**: Double-check all connections for tightness and correctness to prevent any potential short circuits or energy losses.

Wiring Between Charge Controllers and Batteries

The charge controller regulates the flow of electricity from the solar panels to the batteries, ensuring that they are charged efficiently and safely. Here's how to wire between the charge controller and batteries:

1. **Choose the Right Wires**: Use wires with the appropriate gauge to handle the current your system will carry. Larger systems require thicker wires to prevent overheating and power loss.

2. **Connect to Terminals**: Attach the positive and negative wires from the charge controller to the corresponding terminals on the batteries. Ensure secure and correct connections to avoid issues during operation.

3. **Verify Settings**: Configure the charge controller settings according to the battery type and system requirements. This guarantees efficient charging and prolongs battery life.

Connecting Batteries to the Inverter

The inverter converts the DC power from the batteries into AC power for your household appliances. Proper wiring is crucial for safe and efficient operation:

1. **Prepare the Inverter**: Ensure the inverter is turned off before making connections.

2. **Battery Connections**: Connect the positive and negative terminals of the inverter to the corresponding battery terminals. Use heavy-duty cables to handle the high current.

3. **Check for Tightness**: Inspect all connections for security and tightness to avoid potential electrical faults.

Ensuring Proper Wiring for System Efficiency

Efficient wiring is always a good measure to maximizing the performance of your solar power system. To ensure optimal efficiency:

1. **Minimize Voltage Drop**: Use the shortest wire lengths possible and choose wires with adequate thickness to reduce resistance and voltage drop.

2. **Label Wires**: Clearly label all wires to prevent confusion and mistakes during maintenance or troubleshooting.

3. **Organize Cables**: Route cables neatly and securely to prevent damage and reduce the risk of accidental disconnections.

Troubleshooting Wiring Issues

Even with careful installation, you may encounter wiring issues. Here's how to troubleshoot and fix common issues:

Common wiring issues and solutions

Loose Connections: Loose connections can lead to intermittent power issues. Tighten all connections and check for signs of wear or corrosion.

Short Circuits: If you experience short circuits, inspect wiring for damage or exposed wires. Replace any broken portions and ensure appropriate insulation.

Inconsistent Power Output: If your system isn't performing as expected, check for voltage drops and ensure all components are correctly connected and functioning.

Diagnostic Tools and Techniques

To effectively diagnose wiring issues:

1. **Multi meter**: Use a digital multi meter to measure voltage, current, and resistance at various points in your system. This helps identify faults and ensure everything is operating within specifications.

2. **Clamp Meter**: A clamp meter can measure current flow without disconnecting wires, making it useful for live system diagnostics.

3. **Visual Inspection**: Regularly inspect wiring for signs of damage, wear, or loose connections.

Ensuring Proper Functionality and Safety

Proper wiring and overcurrent protection are crucial for the safe and efficient operation of your solar power system. Follow these best practices to ensure functionality and safety:

1. **Overcurrent Protection**: Install fuses or circuit breakers to protect your wiring and components from excessive current. Ensure they are rated for your system's specifications.

2. **Grounding**: Properly ground your system to prevent electrical shocks and reduce the risk of damage due to electrical faults.

3. **Regular Maintenance**: Perform regular checks and maintenance on your wiring and overcurrent protection devices to ensure ongoing safety and performance.

Next

CHAPTER NINE

BUILDING YOUR SOLAR SETUP

This chapter walks you through the necessary processes for designing, sizing, and installing a unique solar setup that is suited to your specific requirements. Whether you're building up a little off-grid hut or a bigger house system, here will offer you the knowledge and confidence to design a functioning and efficient solar power system.

Designing a Custom Solar Setup from Scratch

The first stage in creating your solar system is to design a system that meets your individual requirements. Begin by calculating your energy needs and determining your available space. Consider the following:

- **Energy Needs Assessment**: Calculate your daily energy consumption. List all the appliances and devices you intend to power, noting their wattage and usage hours. This will help you determine the size of your solar array and battery bank.

- **Space Availability**: Evaluate the space where you plan to install your solar panels. Ensure it has adequate sunlight exposure and is free from obstructions like trees or buildings that might cause shading.

- **Budget**: Set a budget for your solar system. Include costs for panels, batteries, charge controllers, inverters, mounting hardware, and installation.

Sizing the Solar Array and Battery Bank

Once you have a clear understanding of your needs, it's time to size your solar array and battery bank:

- **Solar Array Sizing**: Determine the number of solar panels required to meet your energy needs. Consider the panel's wattage, efficiency, and the average peak sun hours in your location. A general formula is to divide your daily energy consumption (in watt-hours) by the number of peak sun hours to get the required panel wattage.

- **Battery Bank Sizing**: Calculate the battery bank size based on your daily energy consumption and desired days of autonomy (the number of days you want your system to run without sunlight). Use deep-cycle batteries and ensure you have enough capacity to handle your needs.

Choosing the Right Solar Panels and Batteries

Selecting high-quality components is crucial for the performance and longevity of your solar system:

- **Solar Panels**: Choose panels with high efficiency and good durability. Mono crystalline panels generally offer higher efficiency and better performance in low-light conditions compared to polycrystalline panels.

- **Batteries**: Opt for deep-cycle batteries that can handle repeated charging and discharging. Lead-acid batteries are common and cost-effective, while lithium-ion batteries offer better efficiency and longer life.

Selecting the Appropriate Charge Controller and Inverter

The charge controller and inverter are critical components that manage and convert the solar power:

- **Charge Controller**: Choose between PWM (Pulse Width Modulation) and MPPT (Maximum Power Point Tracking) controllers. MPPT controllers are more efficient and recommended for larger systems.

- **Inverter**: Select an inverter that matches your system's voltage and power requirements. Pure sine wave inverters are preferred for their stability and compatibility with sensitive electronics.

Step-by-Step Guide to System Installation

Follow these steps to install your solar setup:

1. Preparing the Installation Site

- **Site Assessment**: Ensure the installation area is clean and structurally sound.
- **Permits**: Obtain any necessary permits for your solar installation.

Preparing the Installation Site

Mounting Hardware: Install the mounting brackets and rails securely on your roof or ground mount.

Panel Installation: Attach the solar panels to the mounting hardware, ensuring they are properly aligned and secured.

Mounting Solar Panels

- **Connecting Panels**: Wire the panels in series or parallel based on your system design. Use the appropriate connectors and ensure that all connections are secure.

- **Battery Connections**: Connect the solar array to the battery bank using appropriate wiring and connectors.

Wiring the System: From Panels to Battery

- **Charge Controller**: Connect the charge controller to the battery bank and solar panels. Follow the manufacturer's instructions for wiring and settings.

- **Inverter**: Wire the inverter to the battery bank and your household appliances. Ensure the inverter is properly grounded and configured.

Connecting the Charge Controller and Inverter

Initial Test: Turn on the system and check all connections for proper functionality.

Calibration: Adjust the charge controller and inverter settings to match your system specifications and optimize performance.

System Testing and Calibration

- **Regular Checks**: Monitor the system regularly to ensure its operating efficiently. Check battery levels, panel cleanliness, and wiring integrity.

- **Maintenance**: Perform routine maintenance to keep your system in good working condition. Clean the panels periodically and inspect connections for wear.

Wiring Considerations for Mobile Off-Grid Systems

When it comes to mobile off-grid systems—whether for an RV, boat, or tiny home—wiring is a critical component that requires careful planning. Mobility introduces unique challenges, such as vibration, space constraints, and the need for flexibility.

- **Flexibility and Durability:** Mobile systems often experience more wear and tear than stationary setups, so it's essential to choose wires that can handle movement and resist corrosion. Marine-grade cables, which are tinned to resist rust and oxidation, are an excellent choice for these environments.

- **Routing and Protection:** Wires should be routed in a way that minimizes exposure to sharp edges, heat sources, and areas of potential wear. Use protective conduit and grommets to safeguard wires from damage. In an RV or marine setting, it's vital to ensure that all wiring is secure and won't shift during transit.

- **Connections and Fittings:** All connections must be secure, as loose connections can lead to power loss or even fire hazards. Consider using crimped connections with heat-shrink tubing to create a sealed and reliable joint. For mobile systems, quick-disconnect fittings can also be a valuable addition, allowing for easy disassembly if you need to troubleshoot or replace components.

Marine and RV Wiring Considerations

Marine and RV setups have their own set of specific requirements due to the unique environments in which they operate.

- **Marine Systems:** Saltwater is particularly corrosive, so marine systems demand high-quality materials that can withstand harsh conditions. Use marine-grade wire and connectors, which are specifically designed to resist corrosion. Pay close attention to grounding and bonding practices, as improper grounding can lead to galvanic corrosion, which could damage metal parts of your boat.

- **RV Systems:** In an RV, space is often at a premium, so wiring must be both efficient and compact. Additionally, because RVs typically have multiple power sources (shore power, generator, solar), it's crucial to have a well-designed system that can manage these inputs without overloading circuits. Investing in a high-quality power management system can help ensure that your RV's electrical setup is both safe and efficient.

Managing Voltage Drop and Wire Sizing

Voltage drop can significantly impact the performance of your solar system, especially in mobile setups where wire runs may be longer or exposed to more varied conditions.

- **Understanding Voltage Drop:** Voltage drop occurs when the electrical potential decreases as the current flows through the wire, which can lead to insufficient power reaching your devices or batteries. This is particularly critical in low-voltage systems like those commonly used in RVs and boats, where even a small drop can lead to significant power loss.

- **Proper Wire Sizing:** To minimize voltage drop, it's essential to choose the correct wire size. This is determined by the length of the wire run, the amount of current it needs to carry, and the acceptable voltage drop for your system. Generally, larger wires have less resistance and therefore less voltage drop, but they are also more expensive and harder to route.

- **Tools for Calculation:** Utilize online voltage drop calculators or consult tables that can help you determine the appropriate wire size for your system. It's always better to err on the side of caution with thicker wire, especially in important circuits that power essential components.

Cost Reduction Strategies for Solar Installations

Setting up a solar power system can be a significant investment, but there are strategies you can employ to reduce costs without compromising on quality or efficiency.

Sourcing Affordable Components

Purchasing low-cost components is one of the most efficient strategies to save money. This does not imply purchasing the lowest things available, but rather opting for excellent components at a cheaper cost. Online markets, local wholesalers, and even surplus stores may provide great deals on solar panels, inverters, and batteries. Consider buying used or reconditioned components, especially for long-lasting items like solar panels.

Bulk purchase can also result in cost reductions, especially for high-volume wiring and connections. Another idea is to keep an eye out for specials or discounts provided by manufacturers, particularly around the end of the year when they may be clearing out older inventory.

DIY Installation vs. Professional Installation

Another big cost-saving strategy is to do your own installation. Installing your own solar system may significantly cut labor expenses, which can account for a large amount of the total cost. Step-by-step instructions for a successful installation may be found in a variety of sites, including online tutorials, guidelines, and community forums.

However, DIY is not for everyone. If you are uncomfortable dealing with electrical systems or lack the essential tools and abilities, professional installation may be a safer option. While more expensive initially, expert installation ensures that your system is set up accurately and effectively, which may save you money in the long term by preventing any faults or inefficiencies.

Incentives and Rebates for Solar Power

Do not underestimate the financial benefits offered by solar systems. Many countries provide rebates, tax credits, and other incentives to promote the use of solar energy. These can considerably cut your total expenditures while increasing the return on your investment.

Before making a purchase, research the incentives offered in your region. Some schemes require prior clearance, and the savings might be significant. In certain countries, government tax incentives might cover a significant portion of installation expenses. Furthermore, several locations provide discounts for particular components, such as batteries or charge controllers, significantly lowering the entire cost.

Maintenance and Troubleshooting Tips

Maintaining your solar power system is important to its lifespan and performance. While solar panels are generally low-maintenance, the other components batteries, inverters, and wiring require frequent care to keep everything functioning well.

Common Issues and Solutions

Inspecting Solar Panels:

- Regularly clean your solar panels to remove dust, bird droppings, and debris. A buildup of dirt can significantly reduce their efficiency. A simple rinse with water and a gentle toothbrush should suffice.

- Check for physical damage such as cracks or loose wiring, especially after severe weather conditions. Even minor damage can affect performance.

Regular Maintenance Tasks

Regularly clean your solar panels to remove dust, bird droppings, and debris. A buildup of dirt can significantly reduce their efficiency.

Check for physical damage such as cracks or loose wiring, especially after severe weather conditions. Even minor damage can affect performance.

1. **Battery Health Check**:
   - Periodically inspect your batteries for signs of wear and tear, corrosion, or leakage. Keep terminals clean and free of corrosion by applying a corrosion-resistant coating.

- Monitor the battery charge levels. Over time, batteries may lose their ability to hold a charge, indicating it might be time to replace them.

2. **Inverter and Charge Controller Maintenance**:

    - Ensure your inverter is functioning properly by checking its output regularly. Look for any unusual noises or fluctuations in power output, which could signal a problem.

    - The charge controller should also be monitored to ensure its properly regulating the power flowing into your batteries. Check for error codes or warning lights that could indicate issues.

3. **Wiring and Connections**:

    - Regularly inspect all wiring for signs of wear, loose connections, or exposed wires. Properly secure and insulate any loose or damaged wires to prevent short circuits or power loss.

Common Issues and Solutions

1. **Low Power Output**:

    - **Issue**: If your system is producing less power than expected, it could be due to dirty panels, shading, or wiring issues.

    - **Solution**: Clean the panels, remove any sources of shade, and inspect the wiring for faults or loose connections.

2. **Battery Not Holding Charge**:

    - **Issue**: Batteries that discharge quickly or fail to hold a charge may be old, damaged, or improperly maintained.

    - **Solution**: Test each battery's voltage and replace any that are underperforming. Ensure that your charge controller is functioning correctly to prevent overcharging or deep discharging.

3. **Inverter Failures**:

    o **Issue**: If the inverter isn't working, it could be due to overloading, wiring issues, or internal faults.

    o **Solution**: Reduce the load on the inverter and check all connections. If the problem persists, consult the manufacturer's troubleshooting guide or consider replacing the unit.

4. **System Shutdowns**:

    o **Issue**: Unexpected system shutdowns might be caused by faulty wiring, overheating, or insufficient battery capacity.

    o **Solution**: Inspect all connections, ensure proper ventilation to avoid overheating, and consider adding more batteries to increase capacity.

Expanding Your System: Future-Proofing Your Setup

As your energy needs grow, you may find that your initial solar setup no longer meets your demands. Fortunately, solar power systems are highly scalable. This section will guide you through expanding your system to accommodate increased usage, ensuring it remains efficient and reliable.

Adding More Panels and Batteries

1. **Assessing Your Current Setup**:
    - Before expanding, evaluate your current system's capacity and performance. Determine if you need more energy output (more panels) or increased storage (more batteries).

- Calculate your energy needs by reviewing your power consumption over the past months. This will give you a clear picture of how much additional capacity you require.

2. **Selecting and Installing Additional Panels**:

    - When adding more panels, ensure they are compatible with your existing system. Consider the voltage and current ratings, and choose panels that match or complement your current setup.

    - Plan for the additional space required for new panels, and ensure your mounting structure can support the extra weight.

3. **Expanding Battery Storage**:

    - Adding more batteries increases your system's storage capacity, allowing you to store more energy for use during nighttime or cloudy days.

- Ensure that new batteries are compatible with your existing system and that your charge controller can handle the increased load.

Upgrading Components

Upgrading the Inverter:

As your system grows, your inverter may need to be upgraded to handle higher power loads. Choose an inverter with sufficient capacity and features that support future expansion.

Enhancing the Charge Controller:

If you're adding significant capacity to your system, consider upgrading your charge controller to ensure it can manage the increased power flow and prevent overcharging.

Improving System Efficiency:

Invest in more efficient components, such as higher-quality wiring or a more advanced inverter, to reduce power loss and maximize energy output.

Monitoring and Automation:

Consider integrating monitoring and automation tools to optimize your system. These tools can track performance in real-time, alert you to potential issues, and automatically adjust settings for maximum efficiency.

CHAPTER TEN

MONITORING AND MAINTAINING YOUR SOLAR SYSTEM

Obtaining energy independence with an off-grid solar system is an impressive accomplishment, but the path does not end with installation. To ensure that your system runs efficiently over time, regular monitoring and maintenance are required.

Solar System Performance Monitoring Tools

Monitoring your solar system's performance is critical for recognizing problems early and guaranteeing peak energy production. There are several tools available to assist you keeping an eye on how your system is doing. These instruments vary from basic meters that indicate voltage and current levels to complex software that records every element of your system's output.

Some key monitoring tools include

- **Solar Power Meters:** Simple devices that measure the power output from your solar panels.

- **Energy Monitors:** Provide detailed insights into how much energy your system generates and consumes, often offering data in easy-to-read formats.

- **Battery Monitors:** These devices track the state of charge, voltage, and health of your batteries, giving you a clear picture of their condition.

Types of Monitoring Systems

Monitoring systems can be categorized based on their complexity and the level of detail they provide. Here are some common types:

- **Basic Onsite Monitors:** These systems are often built into your inverter or battery management system. They provide real-time data on your solar power generation and consumption but require you to be physically present to access the information.

- **Advanced Monitors with Data Logging:** These systems store data over time, allowing you to analyze trends in your solar system's performance. They are ideal for detecting gradual changes that may indicate a developing problem.

- **Integrated Monitoring Systems:** These are comprehensive systems that monitor every aspect of your solar setup, including panels, inverters, batteries, and load management. They often come with user-friendly interfaces and detailed reporting features.

Choosing the right monitoring system depends on your specific needs, the complexity of your solar setup, and how hands-on you want to be with your system's maintenance.

Remote Monitoring Options

For those who want to stay informed about their solar system's performance while on the go, remote monitoring options offer the ultimate convenience. These systems connect to your solar setup and transmit data to your smartphone, tablet, or computer, allowing you to check on your system from anywhere with an internet connection.

- **Cloud-Based Monitoring:** These systems upload your solar data to the cloud, where it can be accessed through an app or web portal. This option is ideal for off-grid homes in remote locations, as it eliminates the need for you to be onsite to monitor your system.

- **Smartphone Apps:** Many solar equipment manufacturers offer apps that provide real-time alerts and insights into your system's performance. These apps often allow you to set up notifications for specific conditions, such as low battery levels or sudden drops in power generation.

Remote monitoring not only gives you peace of mind but also allows you to respond quickly to any issues, minimizing downtime and maximizing energy production.

Real-Time Data Analysis

Real-time data analysis is a powerful tool that helps you understand how your solar system is performing at any given moment. By continuously monitoring key metrics like energy production, battery charge levels, and power consumption, you can gain valuable insights into your system's efficiency and make informed decisions about its operation.

- **Energy Production vs. Consumption:** Comparing these two metrics in real time allows you to see if your solar panels are meeting your energy needs. If consumption regularly exceeds production, you may need to consider additional panels or energy-saving measures.

- **Battery Health Monitoring:** Real-time analysis of your battery's state of charge and overall health helps you manage energy storage effectively, ensuring you have enough power when you need it most.

- **System Alerts:** Advanced monitoring systems can analyze data in real time and alert you to any anomalies, such as a sudden drop in energy production, which could indicate a problem with your panels or inverters.

Routine Maintenance Procedures

Routine maintenance is essential to keep your solar system operating smoothly and efficiently. While solar panels and batteries are designed to be low-maintenance, neglecting regular checks can lead to reduced performance and costly repairs down the line.

Here are some key maintenance tasks:

- **Panel Cleaning:** Dust, dirt, and debris can accumulate on your solar panels, reducing their efficiency. Regularly cleaning your panels with water and a soft brush ensures they can absorb the maximum amount of sunlight.

- **Inspection of Connections and Wiring:** Over time, connections can loosen, and wiring can degrade. Periodically inspect all electrical connections and cables to ensure they are secure and in good condition.

- **Battery Maintenance:** Depending on the type of batteries you use, maintenance can vary. Lead-acid batteries, for example, may require regular water topping, while lithium-ion batteries generally need less upkeep. Always adhere to the manufacturer's directions for battery maintenance.

- **Monitoring System Calibration:** Ensure that your monitoring systems are accurately calibrated. This includes checking sensors, meters, and other monitoring equipment for proper operation.

- **Annual Professional Check-Up:** While much of the maintenance can be done on your own, it's advisable to have a professional inspect your system annually. They can perform detailed checks that might be beyond the average DIY enthusiast's capability, ensuring your system is in top shape.

Battery Maintenance and Inspection

Batteries store the energy your panels collect, so keeping them in good condition is vital for reliable power supply:

- **Regular Inspections:** Check your batteries monthly. Look for signs of corrosion on the terminals, swelling, or leakage. Corrosion can be cleaned with a mixture of baking soda and water, but be sure to disconnect the battery before cleaning.

- **Water Levels:** If you're using flooded lead-acid batteries, check the water levels regularly. Top up with distilled water as needed, ensuring the plates are submerged but not overfilled.

- **Temperature Monitoring:** Batteries perform best in moderate temperatures. Extreme heat or cold can reduce their lifespan. Consider insulating your battery bank or using temperature-controlled storage if you live in an area with harsh climates.

Checking and Tightening Connections

Loose or corroded connections can lead to inefficient power transfer or even dangerous electrical faults:

- **Monthly Checks:** Once a month, go over all the connections in your system, including the panels, charge controller, batteries, and inverter. Use a torque wrench to ensure all bolts and screws are tight but be careful not to over tighten, which can damage the components.

- **Look for Wear and Tear:** Inspect wires and cables for signs of wear, such as fraying or cracking. Replace any damaged wires immediately to prevent shorts or fires.

Inverter Maintenance and Checks

The inverter is crucial for converting the DC power from your panels into AC power for your home:

- **Cooling System Check:** Inverters generate heat, so they often have cooling fans or heat sinks. Ensure these are free of dust and debris, and check that the fans are operating properly.

- **Firmware Updates:** Some inverters can be updated to improve performance or fix bugs. Check the manufacturer's website often for any accessible updates.

- **Performance Monitoring:** Most modern inverters have a display or connect to an app that lets you monitor performance. Regularly review these metrics to ensure your inverter is functioning optimally.

Troubleshooting Common Issues

Despite your best efforts, problems can arise. Here's how to deal with some of the most prevalent issues:

Identifying and Fixing Performance Issues

If your system isn't generating as much power as expected, start by reviewing your daily and seasonal performance data:

- **Compare Output:** Compare the actual output to the predicted production for your area and time of year. A big disparity might suggest a problem with one or more of the components.

- **Inspect Components:** Examine all system components, including the panels and the inverter, for signs of damage or dysfunction. If you can't locate the problem, try contacting an expert to do a more thorough study.

Dealing with Panel Shading and Debris

Shading from surrounding trees, buildings, or even bird droppings can dramatically affect the effectiveness of the panels:

- **Regular Trimming:** Keep trees and shrubs trimmed down to avoid throwing shadows on your panels.

- **Remove Debris:** Remove any leaves, snow, and other debris as quickly as possible. Even partial shading on one panel might impair the overall system's performance.

Addressing Battery Problems

Battery difficulties can cause a loss of power or lower system efficiency:

- **Voltage Checks:** Regularly check each battery's voltage. If one battery's voltage is much lower than the others, it may be failing and will need to be replaced.

- **Balancing the Bank:** If you detect different charge levels across your battery bank, you may need to equalize the batteries. This technique includes slightly overcharging the batteries to balance the cells; however, it should only be done according to the manufacturer's guidelines.

Resolving Inverter Faults

If your inverter isn't operating properly, it might turn off the entire system:

- **Check Error Codes:** When something goes wrong with an inverter, it usually displays an error code. Refer to your inverter's handbook to understand these codes and identify the required action.

- **Reset the Inverter:** Sometimes merely resetting the inverter can solve minor problems. Turn off the inverter and wait a few minutes before turning it back on.

- **Consult a Professional:** If the problem persists, it might be caused by an internal malfunction that requires expert repair or replacement.

Extending the Lifespan of Your Solar System

Following the maintenance measures listed above can considerably increase the longevity of your solar system. Regular maintenance not only assures maximum operation, but also saves costly issues down the road:

- **Keep Records:** Keep track of your inspections, maintenance chores, and any repairs you've completed. This allows you to monitor the health of your system over time and may be valuable if you need to see a specialist.

- **Stay Informed:** Technology advances, as do best practices in solar power. Keep up with current technologies and consider updates as they become available to keep your system functioning properly for years to come.

Best Practices for Longevity (H2)

Your solar power system is a big investment that, like any other, must be cared for and monitored to maintain its longevity. Here are some suggested practices for extending the life of your system:

- **Regular Inspections:** Make it a practice to examine your solar panels, wiring, and batteries on a regular basis. Inspect for evidence of wear and tear, corrosion, or damage. Early diagnosis of faults can avoid costly repairs down the future.

- **Cleaning the Panels:** Dust, grime, and debris may build up on your solar panels, lowering their effectiveness. Periodically clean your panels with water and a gentle brush or sponge. Avoid using aggressive chemicals that might harm the surface.

- **Battery Maintenance:** If your system relies on lead-acid batteries, check the water levels and verify they are properly maintained. Keep the battery terminals clean and corrosion-free.

Monitoring the status of charge and preventing deep discharges can help lithium-ion batteries last longer.

- **Monitoring System Performance:** Use a monitoring device to keep track of your solar power production, battery levels, and overall system efficiency. Check the data on a regular basis for any irregularities or decreases in efficiency that might suggest a problem.

Protecting Your System from Environmental Damage

Your solar energy system is exposed to the elements, making it susceptible to external influences. Protecting it from these possible hazards is important for its endurance:

- **Weatherproofing Components:** Make that all of your system's components, including as inverters, charge controllers, and wiring, are waterproof. Protect important electronics from rain, wind, and dust by using outdoor-rated enclosures.

- **Lightning Protection:** Install a lightning protection system, particularly if you reside in a thunderstorm-prone location. This may include grounding rods and surge protectors to protect your electronics from power surges generated by lightning strikes.

- **Animal and Pest Control:** Birds, rats, and insects can harm your system by nesting beneath panels or gnawing on wiring. Implement efforts to prevent pests, such as installing critter guards or rodent-resistant wiring.

- **Shielding from Debris:** In areas with frequent storms, flying debris might endanger your panels. Consider building protective barriers or putting panels in less exposed places without losing solar exposure.

Upgrading and Replacing Components

Over time, components in your solar system may need to be upgraded or replaced. Staying on top of this ensures that your system stays efficient and effective:

- **Upgrading to New Technologies:** Solar technology is always improving. Keep a watch out for developments such as more efficient panels, better batteries, and smarter inverters. Upgrading to current technologies might improve your system's efficiency and output.

- **Replacing Worn Components:** Even with proper care, some components will wear out. Batteries normally have a lifespan of 5-15 years, depending on type, whereas inverters can last 10-15 years. Plan for these replacements to prevent unexpected interruptions.

- **Scaling Up:** As your energy requirements increase, you may wish to upgrade your system by adding more panels or batteries. Ensure that your present system can manage the extra load, or consider updating your inverter and charge controller to suit the increase.

Seasonal Adjustments and Optimization

Solar energy output can fluctuate seasonally owing to changes in daylight hours, solar angle, and meteorological conditions. Adapting your system to these changes is crucial for sustaining peak performance:

Adapting to Seasonal Changes

As the seasons shift, so do your energy requirements and the quantity of sunlight available. In the winter, fewer days and lower sun angles can restrict solar output, but in the summer, surplus energy production may occur. This is How to Adapt:

- **Winter Adjustments:** In colder months, energy demand often increases due to heating needs. Ensure your batteries are fully charged, and consider reducing energy consumption or supplementing with an alternative power source.

- **Summer Adjustments:** During summer, your system might produce more energy than needed. Use this time to charge batteries fully, run energy-intensive tasks, or store excess power for future use.

Optimizing Panel Angles and Orientation

The angle and orientation of your solar panels significantly impact their efficiency. Seasonal adjustments to panel positioning can help maximize energy capture:

- **Panel Angle Adjustment:** During winter, adjust your panels to a steeper angle to capture the lower-angle sunlight. In summer, a flatter angle may be more effective. Use a solar tracker if available, which automatically adjusts the panel angle throughout the day and seasons.

- **Orientation Considerations:** Ensure your panels are oriented to capture the maximum amount of sunlight throughout the year. In most regions, this means facing the panels south (in the northern hemisphere) or north (in the southern hemisphere).

Managing Snow and Ice Build-Up

In snowy regions, snow and ice can accumulate on your panels, blocking sunlight and reducing efficiency. Here's how to manage it:

- **Snow Removal:** Use a soft-bristled broom or a specially designed snow rake to clear snow from your panels. Avoid using metal tools, which might harm the surface.

- **Preventing Ice Damage:** Ice can be more problematic than snow, as it can weigh down panels and cause structural damage. Consider installing heaters under the panels or using an anti-icing coating to prevent ice build-up.

Adjusting for Seasonal Energy Needs

Your energy consumption may fluctuate with the seasons. For instance, you might use more heating in winter and more cooling in summer. Adjust your system to match these needs:

Energy Usage Planning: Monitor your energy consumption throughout the year and adjust your system settings accordingly. In winter, prioritize essential loads to conserve power.

Seasonal Maintenance: Perform maintenance checks at the start of each season to ensure your system is ready to handle the upcoming environmental changes. This could include cleaning panels, checking battery health, and inspecting wiring for wear and tear.

Advanced Monitoring Techniques

To get the most out of your off-grid solar system, consider implementing advanced monitoring techniques:

- **Remote Monitoring:** Use remote monitoring systems to keep track of your solar system from anywhere. These systems can alert you to potential issues before they become critical, allowing for quick intervention.

- **Data Analysis:** Analyze the data from your monitoring system to identify patterns in energy production and consumption. This can help you optimize your system for better performance and efficiency.

- **Predictive Maintenance:** Implement predictive maintenance techniques that use data trends to forecast potential issues, allowing you to address them before they cause a failure.

Using Data Loggers for Detailed Analysis

Data loggers are essential tools for anyone serious about getting the most out of their solar power system. These devices record the performance of your system in real time, capturing crucial data points such as voltage, current, and energy output. By analyzing this data, you can gain insights into how efficiently your system is operating and identify any potential issues before they become major problems.

Imagine this: It's the middle of a scorching summer, and your energy consumption has hit an all-time high. You notice a slight dip in your system's performance but can't pinpoint the cause. With a data logger, you could quickly analyze the historical data and spot that the issue started when a specific panel began underperforming, possibly due to dirt or shading. A quick cleaning or adjustment later, and your system is back to its peak performance.

By regularly reviewing data logger reports, you can make informed decisions about maintenance, catch small issues early, and ensure that your solar system is running as efficiently as possible.

Integrating with Smart Home Systems

In today's digital age, smart home systems are revolutionizing the way we interact with our living spaces. Integrating your solar power system with a smart home network allows you to monitor and control your energy production and consumption with unprecedented ease. Imagine being able to adjust your energy usage based on real-time solar output, or receiving an alert on your phone if a component of your system needs attention.

Smart home integration can also automate processes like diverting excess energy to a battery storage system or adjusting appliance usage to coincide with peak solar production. This level of control not only optimizes the efficiency of your solar system but also enhances your overall energy management, potentially saving you money on your utility bills.

Analyzing Historical Data for Performance Trends

To truly understand how well your solar system is performing, you need to look at the bigger picture. Analyzing historical data helps you spot trends over time, such as seasonal variations in energy production or the gradual decline in efficiency of an aging component. By keeping an eye on these trends, you can make proactive decisions that keep your system running smoothly and efficiently.

For example, if you notice that your system consistently underperforms during the winter months, you might consider adjusting the tilt of your panels or even adding a few more to compensate for the reduced sunlight. Historical data analysis turns raw numbers into actionable insights, ensuring your system's long-term success.

Solar System Upgrades

As your energy needs grow or as technology advances, your solar power system may require upgrades to keep up. Whether it's expanding capacity or enhancing efficiency, upgrading your system can help you get the most out of your investment.

Assessing the Need for System Upgrades

Before jumping into upgrades, it's crucial to assess whether they're necessary. Consider your current and future energy needs, any changes in your household or lifestyle, and the age and performance of your existing system. If your energy usage has increased due to new appliances or an expanding family, or if your system is no longer meeting your needs efficiently, it might be time for an upgrade.

Adding Additional Panels or Batteries

One of the most straightforward upgrades is adding more solar panels or batteries to your system. Adding panels increases your system's energy production, helping you capture more sunlight and generate more power. This can be particularly beneficial if you've recently added energy-intensive appliances or if your system is struggling to keep up during certain times of the year.

Adding additional batteries, on the other hand, allows you to store more energy for later use, increasing your energy independence. This is especially useful in off-grid scenarios or areas prone to power outages. More storage means you can rely on your solar system even when the sun isn't shining.

Upgrading Inverters and Charge Controllers

Inverters and charge controllers are the heart of your solar system, converting and managing the energy produced by your panels. As technology advances, more efficient and capable models become available, offering better performance and reliability. Upgrading to a more advanced inverter or charge controller can improve your system's overall efficiency, reduce energy loss, and enhance your ability to monitor and control your power usage.

For instance, if you're integrating your solar system with a smart home system, a new inverter with better connectivity options might be necessary. Similarly, if you've added more panels, a more robust charge controller may be required to handle the increased power.

Safety Considerations (H2)

Maintaining your solar system is essential to keeping it in top condition, but it must be done with safety as a priority. When performing any maintenance work, always follow these precautions:

- **Shut down the System:** Before beginning any maintenance, make sure to completely shut down the system. This includes turning off the solar panels, disconnecting batteries, and cutting off power to inverters. This step is crucial to prevent accidental electrocution or damage to components.

- **Use Proper Tools and Equipment:** Always use the correct tools for the job. Insulated tools are a must when dealing with electrical components, and wearing protective gear like gloves and safety glasses can protect you from potential hazards.

- **Work in Pairs When Possible:** If the task is complex or requires handling heavy equipment, it's safer to have someone with you. Not only does this provide extra hands for the job, but it also ensures that help is readily available if something goes wrong.

Safety Precautions during Maintenance

Electrical components in your solar system can pose significant risks if not handled properly. Here's how to manage them safely:

- **Avoid Direct Contact:** Never touch live wires or terminals with your bare hands. Even with the system shut down, capacitors can store energy and discharge unexpectedly.

- **Check for Faulty Wiring:** Regularly inspect the wiring for signs of wear, corrosion, or loose connections. Faulty wiring is a leading cause of electrical fires, so address any issues immediately.

- **Label everything:** Each component and connection should be clearly labeled. This not only helps in identifying them quickly during maintenance but also prevents confusion and mistakes.

Handling Electrical Components Safely

Handling electrical components requires careful attention. Always use insulated tools to avoid unintentional short connections. If you're cleaning or adjusting connections make sure your hands and tools are dry. Never bypass safety mechanisms like fuses or circuit breakers; these are in place to protect you and your system from electrical faults.

It's important to be aware of the local weather conditions. Avoid working on your solar system during wet or stormy weather, as moisture increases the risk of electric shock.

Emergency Procedures and Contacts

In the event of an emergency, such as a fire or severe electrical fault, you must have a plan in place. Ensure that everyone in your household knows how to shut down the system quickly and safely. Keep emergency contact numbers, including those of your local fire department, electrician, and solar system installer, easily accessible.

Install a fire extinguisher rated for electrical fires in a convenient location near your solar setup. In case of an electrical fire, never use water to extinguish it—use the fire extinguisher or call emergency services immediately.

Financial Considerations

Beyond safety, maintaining your solar system also involves keeping an eye on the financial aspects. Properly evaluating costs, taking advantage of incentives, and tracking long-term savings will help you maximize the return on your investment.

Evaluating Cost vs. Benefit of Upgrades

As technology advances, you might consider upgrading parts of your solar system to improve efficiency or capacity. Before making any upgrades, it's crucial to evaluate the cost versus the potential benefits. Will the upgrade lead to significant energy savings? How long will the improvement take to pay for itself? For example, newer, more efficient solar panels might generate more power, but the installation costs need to be weighed against the long-term savings.

Consult with a solar energy expert to assess the feasibility of any upgrades. They can provide insights into the expected performance improvements and whether these align with your energy goals and budget.

Incentives and Rebates for System Enhancements

Many regions offer financial incentives or rebates for upgrading or expanding your solar system. These might include tax credits, grants, or subsidies for installing more efficient panels, batteries, or other components.

Research what's available in your area and apply for any incentives that can offset the cost of enhancements. Keeping up to date with these programs can make a significant difference in the overall cost-effectiveness of your solar power system.

Tracking Long-Term Savings and Performance

One of the major benefits of solar power is the potential for long-term savings on energy costs. To ensure you're getting the most out of your system, it's important to track its performance over time. This includes monitoring the amount of energy generated, used, and stored, as well as keeping a record of your energy savings.

Use a performance monitoring system to keep tabs on your solar setup's output. Many modern inverters come with built-in monitoring tools that allow you to check your system's performance through a smartphone app or online portal.

Regularly compare your current energy bills with your bills before installing the system to calculate your savings. This not only helps you assess the system's efficiency but also provides valuable data that can inform decisions about future upgrades or maintenance needs.

CHAPTER ELEVEN

ENERGY STORAGE AND BACKUP SOLUTIONS

Off-grid solar power systems rely on energy storage and backup technologies to function properly. They provide the resilience and dependability required to maintain a consistent power supply, even when the sun isn't shining. We will take you through several energy storage alternatives, integrate backup generators, and arrange for energy autonomy during crises. We'll also investigate the many sorts of energy storage technologies, from the tried-and-true lead-acid and lithium-ion batteries to inventive new alternatives.

Exploring Energy Storage Options

Batteries are the most popular type of energy storage, as they store surplus electricity generated by your solar panels during the day for later usage at night or on cloudy days. However, not all batteries are made equally. Each kind has its own set of benefits, weaknesses, and applications, which we shall discuss in depth in this chapter.

Integrating Backup Generators with Solar Systems

While batteries are necessary, they may not always be sufficient, particularly during prolonged periods of low sunshine. This is when backup generators come in handy. Integrating a generator into your solar power system adds an extra degree of protection, guaranteeing that your power requirements are satisfied even in the most extreme situations. We'll look at how to smoothly integrate a backup generator into your solar system, including the best practices for choosing and sizing a generator to match your setup.

Battery Management Systems (BMS)

A Battery Management System (BMS) is an essential part of any battery storage arrangement. The BMS monitors and maintains your batteries' performance, ensuring they function within safe limits and maximizing their longevity. In this part, we'll dig into the operations of a BMS, including monitoring battery charge levels, balancing cells, preventing overcharging and deep draining, and regulating temperature conditions.

Planning for Energy Autonomy during Emergencies

One of the key reasons many individuals choose off-grid solar power is to obtain energy independence, particularly during crises. Having a dependable energy source can save your life in the case of a natural disaster, a power outage, or any other unanticipated occurrence. This section discusses how to ensure that your energy storage system is resilient enough to manage crises. We'll explore the need of energy audits, redundancy planning, and how to establish an emergency energy plan that keeps you powered regardless of what happens.

Hybrid Systems: Combining Solar with Other Renewable Sources

Solar power is a great source of sustainable energy, but it's not the only option. Combining solar with other renewable sources, such as wind or hydropower, can result in a hybrid system that increases your energy independence. We will talk about the advantages of hybrid systems, how to incorporate several renewable sources, and how to balance and manage these many power inputs. Hybrid systems can provide more constant energy, particularly under varying environmental circumstances.

Types of Energy Storage Technologies

The energy storage industry is quickly growing, with a wide range of technologies available to fulfill various requirements. In this part, we'll look at the most prevalent energy storage devices utilized in off-grid solar systems.

Lead-Acid Batteries: Pros and Cons

Lead-acid batteries are among the oldest and most used forms of energy storage batteries. Lead-acid batteries, known for their dependability and low cost, have long been used in solar power systems. However, there are certain drawbacks, such as a shorter lifespan and reduced efficiency when compared to newer technology. We'll talk about the benefits and drawbacks of lead-acid batteries, and when they might be the best option for your system.

Lithium-Ion Batteries: Pros and Cons

Lithium-ion batteries have grown in popularity in recent years due to their high energy density, extended lifespan, and ability to charge quickly. These batteries are more efficient and have a higher depth of discharge than lead-acid batteries, making them the preferred option for many off-grid solar systems. However, they are more costly and require careful maintenance. This section investigates the benefits and drawbacks of lithium-ion batteries and how they compare to other alternatives.

Flow Batteries: An Overview

Flow batteries are a new technology that promises to provide long-term energy storage while also being very scalable. Unlike traditional batteries, flow batteries store energy in liquid electrolytes that may be refreshed to increase battery capacity. This section gives an explanation of how flow batteries function, their potential benefits, and the present status of this technology in the off-grid solar industry.

Emerging Energy Storage Technologies

The subject of energy storage is constantly evolving, with new technologies emerging that potentially transform how we store and consume energy. From solid-state batteries to sophisticated super capacitors and beyond, this section focuses on the most promising developing technologies that might determine the future of off-grid solar power. We'll talk about their possible uses, advantages, and the obstacles they face before becoming popular.

Maintenance of Energy Storage Systems

Energy storage solutions, particularly batteries, are the foundation of any off-grid solar system. Proper maintenance is essential for ensuring they run smoothly and last as long as possible. Neglecting maintenance can result in lower performance, costlier replacements, and even system breakdowns. Here's how to keep your energy storage systems in great shape.

Regular Inspection and Testing

Regular inspection and testing are key components of keeping your energy storage system in good working order. Your batteries, much like your car's oil, need to be checked on a regular basis to guarantee they're working properly.

- **Visual Inspections:** Begin with a visual evaluation of your batteries. Look for corrosion on the terminals, swelling in the battery shell, and leaks. These might be early warning signs of impending issues that require rapid action.

- **Voltage and Specific Gravity Testing:** Test the voltage of each battery in your system on a regular basis to verify that they are properly charged. For flooded lead-acid batteries, specific gravity measurements with a hydrometer can offer information into the level of charge and health of the battery cells.

- **Load Testing:** Load test your batteries on a regular basis to see how well they will function under real-world settings. This helps discover weak or failing batteries before they create more serious concerns.

Ensuring Proper Ventilation and Temperature Control

Batteries can be sensitive to their surroundings, particularly temperature. Proper ventilation and temperature management are vital for preserving the durability and safety of your energy storage system.

- **Ventilation:** Batteries, especially those that are flooded or vented, require appropriate ventilation to prevent the accumulation of gasses such as hydrogen, which can be harmful. Make sure your battery bank is located in a well-ventilated area with enough airflow to disperse heat and gasses.

- **Temperature Control:** Extreme temperatures have a substantial impact on battery performance and longevity. High temperatures can cause corrosion and water loss in flooded batteries, whereas low temperatures can impair the battery's capacity to supply power. Maintain a steady temperature range for your batteries, ideally between 50°F and 80°F (10°C and 27°C).

Extending Battery Life through Proper Usage

Proper usage habits can extend the life of your batteries, saving you money and ensuring your system runs smoothly for years come.

- **Avoid Deep Discharges:** Repeatedly draining your batteries to very low levels can drastically shorten their lifespan. Try to avoid discharges below 50% of the battery's capacity, and use a charge controller with low-voltage disconnects features to prevent accidental over-discharge.

- **Equalization Charging:** For flooded lead-acid batteries, periodic equalization charging can help balance the charge between cells, prevent sulfation, and extend battery life. This process involves overcharging the battery slightly to equalize the voltage across all cells.

- **Routine Maintenance:** Regularly check the water levels in flooded batteries and top them up with distilled water as needed. Prevent corrosion by cleaning the battery terminals and applying a protective coating.

Backup Power Solutions

Even with a well-maintained energy storage system, there may be times when you will need additional backup power. Whether it's due to extended periods of low sunlight or unexpected system failures, having a reliable backup solution is important.

Grid-Tied vs. Off-Grid Backup Systems

Choosing between a grid-tied and an off-grid backup system depends on your location, energy needs, and goals.

- **Grid-Tied Backup Systems:** If you're in an area with a reliable grid connection, a grid-tied system can serve as your primary backup. These systems allow you to draw power from the grid when your solar battery storage is insufficient. Additionally, if you generate excess solar power, you can often sell it back to the grid, providing a financial benefit.

- **Off-Grid Backup Systems:** For those in remote locations or who want complete energy independence, off-grid backup systems are the way to go. These systems rely on additional energy storage, generators, or other renewable sources like wind or hydro to provide backup power when needed. Off-grid systems are ideal for those seeking resilience against grid outages and who want to reduce their reliance on external power sources.

Automatic Transfer Switches (ATS)

An automatic transfer switch (ATS) is an essential component of a backup power system. It ensures a seamless transition between your primary power source and your backup power source, such as a generator or the grid.

- **How ATS Works:** When your primary power source fails, the ATS automatically switches to the backup source, ensuring uninterrupted power supply. Once the primary source is restored, the ATS will switch back, minimizing disruption.

- **Installation Considerations:** ATS installation should be done by a professional to ensure its correctly integrated with your solar and battery systems. Proper installation will prevent damage to your equipment and ensure the safe and efficient operation of your backup power.

Inverter/Charger Integration with Backup Generators

Inverter/charger units play a critical role in integrating backup generators into your off-grid system. These devices not only convert DC power from your batteries to AC power for your home but also manage charging your batteries when using a generator.

- **Seamless Integration:** A well-designed inverter/charger system can automatically start your backup generator when battery levels get too low. It will then manage the charging process, ensuring that your batteries are charged efficiently and safely.

- **Efficiency Considerations:** Look for inverter/chargers that offer high efficiency and low idle power consumption. This will maximize the energy you store and minimize waste, helping you get the most out of your generator and battery system.

Energy Storage Safety Considerations

When it comes to energy storage systems, safety is crucial. Batteries, particularly big off-grid systems, may store a huge quantity of energy, which, if misused, might result in hazardous circumstances. Understanding the safety factors is the first step toward ensuring your system is not only functional but also secure.

Battery Safety Guidelines

Proper Ventilation: Always make sure the battery storage room is adequately aired. Batteries, particularly lead-acid ones, can release hazardous and combustible gasses. Proper ventilation disperses these gasses and decreases the risk of explosion.

Temperature Control: Batteries are vulnerable to temperature extremes. Too much heat can cause thermal runaway, a condition in which the battery temperature rises quickly and unpredictably, potentially resulting in a fire. Extreme cold temperatures, on the other hand, can degrade battery performance. Store your batteries in a temperature-controlled location whenever feasible.

Secure Installation: Ensure that batteries are securely installed in an upright position and placed on non-conductive, stable surfaces. This minimizes the risk of short circuits or accidental damage.

Regular Inspection: Regularly check your batteries for signs of wear, corrosion, or damage. Early discovery of possible difficulties can help to avoid larger problems in the future.

Fire Protection and Risk Mitigation

1. **Fire Suppression Systems:** Install a fire suppression system in the battery storage area. Fire extinguishers rated for electrical fires should be readily accessible, and, if possible, automated fire suppression systems like sprinklers or gas-based systems should be considered.

2. **Risk Mitigation through Design:** Design your energy storage system with safety in mind. This includes spacing batteries adequately, using insulated cables, and installing proper fusing and

disconnects to quickly shut down the system in case of an emergency.

3. **Emergency Procedures:** Establish clear emergency procedures. Every person using or maintaining the system should know how to shut it down quickly and safely if needed.

Compliance with National Electric Codes (NEC)

Compliance with the National Electric Codes (NEC) is not just about following the law—it's about ensuring your system is as safe and reliable as possible. The NEC provides detailed guidelines on the installation and maintenance of energy storage systems, including:

1. **Proper Sizing and Installation:** The NEC outlines how to properly size and install batteries, inverters, and other components to prevent overloading and overheating.

2. **Grounding:** Proper grounding of your system is essential for safety. The NEC details the

requirements for grounding energy storage systems to prevent electric shock and reduce the risk of fire.

3. **Labeling and Signage:** Clear labeling of components, especially disconnects and breakers, is crucial for anyone who may need to interact with the system, particularly in an emergency.

Monitoring and Optimizing Energy Storage

Once your energy storage system is safely installed, the next step is to monitor and optimize its performance. Effective monitoring helps ensure that your batteries are charging and discharging correctly, and it can alert you to any issues before they become serious problems.

Remote Monitoring Systems

1. **Real-Time Data:** Remote monitoring systems provide real-time data on your battery's performance, including voltage, current, temperature, and state of charge. This data can be

accessed from anywhere, giving you the flexibility to monitor your system even when you're away.

2. **Alerts and Notifications:** Many remote monitoring systems can be set to alert you via text or email if there are any anomalies, such as low battery voltage or high temperature. This allows for quick intervention, preventing potential issues from escalating.

3. **User-Friendly Interfaces:** Modern remote monitoring systems come with user-friendly interfaces that allow even non-experts to understand the performance of their energy storage system.

Data Analytics for Optimizing Performance

1. **Usage Patterns:** By analyzing your energy usage patterns, you can optimize how and when you use stored energy, ensuring maximum efficiency and longevity of your batteries.

2. **Predictive Maintenance:** Data analytics can help predict when maintenance might be needed. For example, if a battery's performance is degrading faster than expected, the system can alert you to check for potential issues.

3. **Performance Benchmarking:** Compare your system's performance against benchmarks to ensure it's operating at peak efficiency. This can help you identify areas for improvement, such as adjusting your energy usage habits or upgrading components.

Troubleshooting Common Energy Storage Issues

Despite your best efforts, issues with energy storage systems can arise. Being prepared to troubleshoot these issues can save time, money, and frustration.

1. **Battery Not Holding Charge:** If your battery isn't holding a charge, the problem could be due to sulfation (in lead-acid batteries), poor connections, or even a faulty battery. Start by checking the

connections and testing the battery with a voltmeter.

2. **Overheating:** If your battery is overheating, it's essential to address this immediately. Overheating could be caused by excessive charging, poor ventilation, or a failing battery. Ensure that your charging parameters are correct and that the battery is in a well-ventilated area.

3. **Inconsistent Power Output:** Inconsistent power output may result from issues with the inverter, poor battery health, or even environmental factors like temperature. Check the inverter settings and perform a battery health check to identify the root cause.

4. **Error Codes on Monitoring System:** Modern energy storage systems often come with monitoring systems that display error codes when something is wrong. Familiarize yourself with these codes and what they mean. Refer to your

system's manual or contact technical support if you encounter an unfamiliar code.

Next

CHAPTER TWELVE

FINANCIAL CONSIDERATIONS AND INCENTIVES

Investing in off-grid solar power is not just an ecologically ethical option, but also a big financial one. Understanding the economic sides of solar energy will help you maximize your investment, allowing you to save money while also taking advantage of any incentives. We'll lead you through the most important financial issues, such as calculating the return on investment (ROI) and researching various funding choices and incentives.

Calculating the ROI of Your Solar Investment

When investing in an off-grid solar system, you must consider the return on investment (ROI). This statistic will help you determine how long it will take for your solar system to pay for itself in the form of lower power costs.

To determine ROI, first estimate your overall system cost, which includes equipment, installation, and maintenance. Next, calculate your annual energy savings—the amount you would have paid your utility provider if you had not used solar power. Divide the total system cost by your yearly savings to calculate the payback time. The shorter the payback period, the greater you're ROI.

Example: Let's say your solar system costs $20,000, and you save $2,500 annually on electricity. Your payback period would be $20,000 ÷ $2,500 = 8 years. After this period, your solar system essentially pays for itself, and every year after that contributes to a positive ROI.

Understanding Tax Credits and Incentives

Governments recognize the value of solar energy and often provide financial incentives to encourage adoption. These incentives can significantly reduce the upfront cost of your solar system, making it more affordable.

Federal Tax Credits

The federal government offers a Solar Investment Tax Credit (ITC) that allows you to deduct a portion of your solar system costs from your federal taxes. As of the latest policy, you can deduct 30% of the cost of installing a solar energy system from your federal taxes. This credit applies to both residential and commercial systems and can be claimed whether you purchase the system outright or finance it.

State and Local Incentives

Many state and municipal governments provide their own incentives. These might include tax breaks, rebates, and grants. These can include tax breaks, refunds, and grants. For instance, some states offer a state tax credit that reduces your state tax liability, while others provide rebates that lower the upfront cost of your system.

Tip: Check your state's energy office or use online tools like the Database of State Incentives for Renewables &

Efficiency (DSIRE) to find out what incentives are available in your area.

Solar Renewable Energy Certificates (SRECs)

In some states, solar system owners can earn Solar Renewable Energy Certificates (SRECs) for the electricity their systems produce. These certificates can be sold to utility companies, which need them to meet renewable energy requirements. The revenue from selling SRECs can provide a significant financial return, making your solar investment even more profitable.

Financing Options for Solar Installations

If the initial cost of solar installation is onerous, many financing alternatives might help make it more reasonable. Understanding these options will help you choose the one that best fits your financial situation.

Loans: Secured vs. Unsecured

Secured loans are backed by collateral, such your home. Because they're less risky for lenders, secured loans often have lower interest rates. However, if you default on the loan, the lender could take possession of the collateral.

Unsecured Loans: These loans don't require collateral but typically come with higher interest rates. They can be easier to obtain if you don't want to risk your assets, but they may end up costing more in the long run.

Leasing vs. Purchasing

- **Leasing**: Leasing allows you to use a solar system owned by a third party. You pay a monthly fee to use the system and benefit from the electricity it produces. The advantage of leasing is the lower upfront cost, but you won't be eligible for most tax credits or incentives.

- **Purchasing**: Purchasing a solar system outright gives you full ownership and access to all available incentives. While the initial cost is higher, the long-term savings and potential ROI are also greater. Additionally, owning the system increases the value of your property.

Power Purchase Agreements (PPAs)

A Power Purchase Agreement (PPA) is another financing option where a solar provider installs and maintains a solar system on your property, and you purchase the electricity it generates at a fixed rate. PPAs typically have little to no upfront cost, and the electricity rate is often lower than that of your utility company. However, like leasing, you won't own the system or benefit directly from incentives.

Long-Term Savings and Environmental Impact

One of the most compelling reasons to invest in off-grid solar power is the potential for significant long-term savings. While the upfront costs might seem daunting, the savings on energy bills over the years can be substantial. Unlike traditional power sources that require ongoing payments to utility companies, solar energy allows you to harness the power of the sun—a free and renewable resource. Over time, these savings add up, making solar power a financially savvy choice.

But the benefits aren't just financial. Off-grid solar power also offers profound environmental advantages. By generating your electricity, you reduce your dependence on fossil fuels, which are a major contributor to greenhouse gas emissions. This reduction in carbon footprint not only helps combat climate change but also contributes to a cleaner, healthier planet for future generations.

Energy Cost Savings over Time

Solar power offers a way to lock in your energy costs for the long haul. Unlike traditional energy sources, where prices fluctuate based on market conditions, solar energy provides a stable and predictable cost structure. Once your system is installed, the ongoing costs are minimal—primarily limited to maintenance. This predictability allows you to plan your finances more effectively, knowing that your energy costs won't unexpectedly skyrocket.

Over time, the cumulative savings from reduced or eliminated energy bills can be significant. In many cases, the money saved on energy costs can exceed the initial investment in just a few years, making solar power a financially attractive option.

Carbon Footprint Reduction

The environmental impact of off-grid solar power is one of its most compelling benefits. Traditional energy sources, such as coal and natural gas, release significant amounts of carbon dioxide and other harmful pollutants into the atmosphere. These emissions contribute to climate change and have adverse effects on public health.

Switching to solar electricity will dramatically lower your carbon impact. Solar panels create power without generating greenhouse gases; therefore they are a clean and sustainable energy source. Over the lifespan of your solar system, the reduction in carbon emissions can be equivalent to planting thousands of trees or removing several cars from the road.

Evaluating the Total Cost of Ownership

When considering an off-grid solar system, it's essential to evaluate the total cost of ownership (TCO). This includes not just the initial installation costs but also ongoing expenses such as maintenance, repairs, and insurance. Understanding the TCO will give you a more accurate picture of the financial commitment involved and help you plan your investment accordingly.

Initial Installation Costs

The initial cost of setting up an off-grid solar system can be significant, encompassing the purchase of solar panels, inverters, batteries, and other necessary equipment. Additionally, there are costs associated with the installation process, such as labor and permits. While these upfront costs might seem high, it's important to view them as a long-term investment that will pay off in the years to come.

Maintenance and Repair Costs

While solar systems are generally low-maintenance, they are not maintenance-free. Regular inspections and occasional repairs are necessary to ensure your system operates efficiently. Over time, batteries may need to be replaced, and inverters might require servicing. These costs should be factored into your long-term financial planning to avoid unexpected expenses.

Insurance Considerations

Protecting your investment in solar power is important. Many homeowners opt to include their solar system in their home insurance policy, but this can lead to increased premiums. It's important to discuss your options with your insurance provider to understand the coverage available and any additional costs that might be associated with insuring your system.

Payback Period Analysis

The payback period is the time it takes for your energy savings to equal the initial investment in your solar system. Understanding this period is a point to evaluating the financial viability of your off-grid solar project. A shorter payback period means you'll start seeing a return on your investment sooner, while a longer period indicates a slower return.

Factors Influencing Payback Period

Several factors can influence the payback period, including the cost of the solar system, the amount of energy you consume, local energy prices, and any available incentives or rebates. Additionally, the efficiency of your solar panels and the amount of sunlight your location receives will play a crucial role in determining how quickly your investment pays off.

Break-Even Point

The break-even point is when the total savings from your solar system equal the total costs. After this point, the energy savings become pure profit. Reaching the break-even point can vary widely depending on the factors mentioned above. However, once you pass this point, every kilowatt-hour of electricity your system generates is essentially free, leading to ongoing financial benefits for years to come.

Impact of Solar on Property Value

Investing in solar power doesn't just reduce your electricity bills; it can also significantly increase the value of your property. Homebuyers are increasingly aware of the benefits of solar energy, from lower utility bills to environmental impact. Studies have shown that homes equipped with solar systems can sell for more than similar homes without them. The value increase often outweighs the initial installation costs, making solar a sound financial investment.

In many regions, properties with solar installations spend less time on the market, appealing to eco-conscious buyers and those looking for long-term savings. Solar power is more than just a green choice; it's a wise financial decision that can make your property stand out in a competitive real estate market.

Real Estate Market Trends

The real estate market is evolving, with sustainability becoming a key selling point. As awareness of climate change and energy efficiency grows, more homebuyers are prioritizing eco-friendly features. Solar power systems are no longer a niche market but a mainstream demand.

Market trends indicate that homes with solar systems are perceived as forward-thinking and cost-efficient, making them more attractive to potential buyers. Moreover, as energy costs continue to rise, the appeal of homes with low or no electricity bills becomes even more compelling. Investing in solar power aligns with these trends, future-proofing your home in a market that increasingly values sustainability.

Resale Value with Solar Systems

When it comes to resale, homes with solar panels often command higher prices. This is partly due to the long-term financial benefits that solar systems offer, such as reduced energy bills and potential earnings from net metering. Homebuyers recognize that solar panels are an asset that will continue to provide returns, making them willing to pay a premium.

Moreover, as energy efficiency becomes more desirable, homes with solar systems are likely to see a faster resale. The ability to market a home as "energy-independent" or "net-zero" can be a significant advantage in today's real estate market, potentially leading to multiple offers and a quicker sale.

Navigating Utility Policies and Rate Structures

Understanding your utility company's policies and rate structures is good when transitioning to solar power. Different utilities offer varying incentives, rates, and policies that can significantly impact the financial benefits of your solar system. Some utilities may offer favorable rates for solar energy, while others might have policies that are less supportive.

It's essential to research and understand these policies before installing your system. For example, some utilities have tiered rate structures, meaning that the more energy you use, the higher your rate. Solar power can help you avoid the highest tiers, saving you money. However, some utilities may have fees or charges for grid access that could offset some savings.

Net Metering Policies

Net metering is one of the most significant incentives for solar power users. It allows you to sell excess energy generated by your solar system back to the grid, often at the same rate that you pay for electricity. This can drastically reduce or even eliminate your electricity bills, making solar a highly attractive investment.

However, net metering policies vary widely depending on your location and utility company. Some areas offer full retail rate net metering, while others might offer reduced rates or limit the amount of energy you can sell back. Understanding your local net metering policy is crucial to maximizing the financial benefits of your solar system.

Time-of-Use Rates

Time-of-use (TOU) rates are another important factor to consider. These rates vary depending on the time of day, with higher rates during peak demand periods. Solar power can be particularly effective in offsetting these higher costs, especially if you can store energy in batteries for use during peak times.

By aligning your energy use with your solar generation and the TOU rate structure, you can maximize your savings. For example, using appliances during off-peak hours when rates are lower can further reduce your energy costs. Understanding TOU rates and adjusting your energy habits accordingly is a smart way to enhance the financial returns of your solar investment.

Utility Rate Changes and Their Impact

Utility rates are not static; they can change due to a variety of factors, including regulatory changes, fuel costs, and infrastructure upgrades. As rates increase, the savings from your solar system become more significant, improving the return on your investment. Conversely, if rates decrease, the financial benefits of solar might not be as pronounced, though this is less common.

It's important to stay informed about potential rate changes and how they might impact your solar savings. Being proactive in understanding and responding to these changes can help you maintain the financial advantages of your solar system over the long term.

Understanding the Impact of Inflation on Solar Investments

Inflation affects everything, including energy costs. As the cost of traditional energy sources rises, the relative value of your solar investment increases. Solar power can act as a hedge against inflation by providing a stable and predictable source of energy, independent of market fluctuations.

With inflation, the cost of electricity from the grid is likely to rise over time, making your solar system more valuable with each passing year. The initial investment in solar can protect you from the uncertainties of future energy prices, offering peace of mind and financial security.

Inflation-Proofing Your Investment

To fully inflation-proof your solar investment, consider options like battery storage or larger systems that can cover more of your energy needs. By generating and storing your own power, you can minimize or eliminate your reliance on the grid, making your energy costs more predictable and stable.

Additionally, locking in your solar installation costs today protects you from the rising costs of both energy and equipment. As inflation drives up prices, the value of your solar investment will continue to grow, providing a buffer against the rising cost of living.

Projecting Future Energy Costs

When considering solar power, it's essential to project future energy costs to understand the long-term savings. While current rates provide a baseline, it's likely that energy costs will rise due to inflation, regulatory changes, and increasing demand. By installing solar, you're essentially locking in today's energy costs for the future.

To project your savings, consider your current energy use, expected rate increases, and the potential for increased energy consumption over time. This forward-looking approach will give you a clearer picture of the financial benefits of solar and help you make an informed decision about your investment.

Maximizing Incentives and Rebates

One of the most powerful ways to reduce the initial cost of your solar setup is by taking advantage of incentives and rebates. Governments and utility companies often offer financial incentives to encourage the adoption of renewable energy. These can come in the form of tax credits, cash rebates, or performance-based incentives.

1. Federal Tax Credits: In many countries, including the United States, federal tax credits allow you to deduct a percentage of your solar installation costs from your taxes. This can significantly lower the overall expense of your solar system.

2. State and Local Incentives: Beyond federal benefits, state and local governments may offer additional rebates or tax incentives. These might vary greatly based on your region; therefore it is critical to study what is available in your area.

3. Utility Company Rebates: Some utility companies provide rebates for solar installations, reducing your upfront costs even further. These rebates are often based on the size of your system or the amount of energy it generates.

Application Processes

To access these incentives and rebates, you'll need to navigate the application processes, which can sometimes be complex. Here's how to streamline the process:

1. Gather Documentation: Before applying, ensure you have all necessary documentation, including proof of purchase, installation contracts, and detailed system specifications.

2. Follow Guidelines: Each incentive program has specific guidelines and requirements. Be sure to read these carefully to avoid any delays or disqualification.

3. Work with Professionals: Consider working with a solar installer or financial advisor who is familiar with the application processes. They can guide you through the steps and help you avoid common pitfalls.

Compliance Requirements

Maximizing incentives and rebates often requires strict adherence to compliance requirements. These may include:

1. System Certification: Ensure that your solar system is certified and meets the technical standards required by the incentive programs.

2. Installation by Certified Professionals: Some incentives require that the system be installed by certified professionals. Verify that your installer has the necessary certifications.

3. Energy Efficiency Standards: Some rebates are contingent on your home or building meeting specific energy efficiency standards. Consider having an energy audit performed to identify any upgrades needed.

Deadlines and Timeframes

Timing is also vital when it comes to financial incentives. Each program has its own deadlines and timeframes for application and approval. Missing these deadlines can mean missing out on valuable financial support.

1. Mark Your Calendar: As soon as you decide to go solar, start tracking the deadlines for each incentive program. Set reminders well in advance to ensure you don't miss any important dates.

2. Apply Early: Programs may have limited funds available on a first-come, first-served basis. Applying early can increase your chances of securing the incentives.

3. Monitor Updates: Incentive programs can change over time. Keep an eye out for any modifications or changes to the programs you are interested in.

Alternative Funding Options

If the initial cost of going solar is a barrier, there are alternative funding options available to help you get started.

1. Solar Loans: Many financial institutions offer loans specifically designed for solar installations. These loans often have lower interest rates and favorable terms compared to traditional loans.

2. Leasing Options: Solar leasing allows you to install a solar system with little to no upfront cost. Instead, you make monthly payments to lease the equipment, often with the option to purchase the system later.

3. Power Purchase Agreements (PPAs): Under a PPA, a third party installs the solar system on your property and you agree to purchase the power generated at a fixed rate. This can be a cost-effective way to go solar without the upfront investment.

Community Solar Programs

For those who can't install solar panels on their own property, community solar programs offer an alternative. These programs allow multiple individuals or households to invest in a shared solar project and receive a portion of the energy generated.

1. Understand the Benefits: Community solar allows you to benefit from solar power without the need to install panels on your property. It's ideal for renters or those with shaded roofs.

2. Research Local Programs: Look for community solar programs in your area. These programs can vary in structure, so it's important to understand the terms before investing.

3. Join Forces: Consider partnering with neighbors or local organizations to participate in a community solar project.

Next

CHAPTER THIRTEEN

CASE STUDIES AND REAL-WORLD EXAMPLES

The world of off-grid solar power is full of inspiring stories of creativity, tenacity, and triumph. This chapter delves into real-world examples that demonstrate the potential of solar energy in a variety of circumstances. These case studies not only illustrate successful installations, but they also provide useful insights into other people's experiences. Whether you're planning a modest DIY project or a large-scale display, these stories will inspire and assist you on your path.

Successful Off-Grid Solar Installations

A family in Colorado's high mountains provided one of the most striking instances of a successful off-grid solar system. Faced with the problem of powering their home away from a utility grid, they devised and constructed a powerful solar system capable of supplying their year-round requirements. They optimized their solar gain by strategically arranging panels to catch sunlight even during the cold months. Their success stands as a tribute to the potential of off-grid solar, showing that, with the appropriate technique, it's possible to live happily and sustainably off the grid.

Lessons Learned from DIY Solar Projects

While professional installations are usually more efficient, many people enjoy planning and installing their own solar systems. Consider a DIY enthusiast in Arizona who took on the challenge of installing an off-grid system for his desert cottage. Through trial and error, he realized the significance of optimum battery size and ventilation—two factors he had previously overlooked. His narrative is a poignant reminder that, while DIY projects may be rewarding, they require rigorous research and a willingness to learn from failures.

Scaling Solar Systems for Larger Properties

As interest in off-grid living develops, so does the demand for scalable solar solutions. A ranch in Texas is an excellent example of how to scale a solar system to power a vast acreage. Beginning with a small array to power basic appliances, the ranch owners progressively extended their system to incorporate more panels and batteries. Today, their solution powers everything from irrigation pumps to refrigeration equipment, demonstrating how solar can expand with your demands and provide reliable power even for large enterprises.

Innovative Uses of Solar Power in Remote Locations

In distant areas where traditional energy sources are limited, solar power has proven to be a game changer. Consider a research facility in the Arctic, where solar panels supply vital electricity throughout the long summer days. Despite the harsh circumstances, the solar array works well, due to improvements such as tilt able panels and anti-freeze coatings. This instance demonstrates how solar technology may be tailored to suit the specific constraints of distant and severe regions.

Adapting Solar Solutions to Unique Environments

Off-grid solar systems must frequently be customized to the unique environmental circumstances in which they operate. To optimize sunshine exposure in the Pacific Northwest's lush woods, one off-grid community has adapted by installing ground-mounted solar arrays in clearings rather than rooftop installations. Their approach emphasizes the necessity of knowing and working with the natural environment to enhance solar energy capture.

Case Study: Off-Grid Living in Harsh Climates

Living off the grid in a severe climate has unique obstacles, as exemplified by a couple living in the Australian Outback. With temperatures frequently surpassing 100°F (38°C) and a scarcity of reliable water sources, they need a solar system that could survive intense heat while providing steady electricity for both living and water pumping. Their approach included bigger solar panels to accommodate for heat-related efficiency losses and a well-insulated battery bank to prevent thermal deterioration. Their experience demonstrates the importance of climatic factors in off-grid solar systems.

Solar Integration in Agricultural Settings

Solar electricity is not just for houses; it is also making its way into agricultural settings. A farm in California's Central Valley uses solar energy to power its irrigation system, which reduces reliance on diesel generators and lowers operating expenses. The use of solar electricity has not only offered financial benefits, but it has also helped the farm meet its environmental goals by lowering its carbon impact. This study highlights the developing synergy between solar energy and agriculture, and how renewable energy can support and boost food production.

Community Solar Projects: Shared Benefits

In certain areas, off-grid solar is a collective effort rather than an individual one. In a distant town in India, a community solar project has altered daily life. The project included the installation of a solar-powered micro grid that supplies energy to all of the village's houses, replacing kerosene lights and diesel generators. The shared benefits are obvious: cleaner air, longer working hours for local companies and higher educational outcomes for youngsters who can now study after dark. This case study demonstrates how community-driven solar projects may benefit whole communities.

Urban Off-Grid Systems: Challenges and Solutions

Off-grid life is commonly associated with rural places, but it is also doable in cities, albeit with distinct problems. A modest apartment complex in New York City has successfully gone off-grid, utilizing solar panels and battery storage to cover its energy requirements. The key to their success was to overcome space limits by using unique solar panel designs and optimizing rooftop area. This urban example proves that with innovation and persistence, off-grid life can be accomplished even in the center of a bustling metropolis.

Mobile Solar Setups: RVs, Boats, and Tiny Homes

Mobile solar systems allow folks who live on the road to travel without losing their comforts at home. One couple's voyage across the United States in their solar-powered RV exemplifies the potential of mobile solar energy. Their solution, which consists of roof-mounted panels and a portable battery bank, can power everything from lights to a small refrigerator, even in distant locations. Similarly, solar installations on boats and tiny dwellings have enabled people to live sustainably while traveling the world.

Solar-Powered Micro grids in Developing Regions

Solar-powered micro grids are having a big influence in poor countries where electricity availability is typically limited. A solar micro grid in rural Kenya has provided energy to a hamlet for the first time. The system powers houses, schools, and a health facility, significantly increasing people' quality of life. This case study demonstrates the transformational potential of solar electricity in locations where traditional infrastructure is limited, as well as its role in promoting economic growth and improving health outcomes.

Emerging Trends in Off-Grid Solar Applications

The off-grid solar environment is constantly changing, as new ideas and technology emerge. The future of off-grid solar is full with exciting possibilities, ranging from better battery storage systems to the incorporation of smart technology that optimizes energy use. While reading through the case studies in this chapter, examine how these developing themes may relate to your own projects and what the future holds for off-grid solar power.

CHAPTER FOURTEEN

THE FUTURE OF SOLAR TECHNOLOGY

The future of solar electricity is in the development of more efficient, cost-effective, and adaptable solar cells. Traditional silicon-based solar panels controlled the market for decades, but they are progressively being replaced by more modern alternatives that promise improved efficiency and cheaper production costs.

Perovskite Solar Cells: Potential and Challenges

Perovskite solar cells are one of the solar industry's most promising technologies. These cells, named after the mineral perovskite, have a distinct crystalline structure that can be tuned to absorb a wide range of the sunlight spectrum. This makes them highly efficient and versatile to a variety of applications, ranging from flexible solar panels to transparent solar cells that may be incorporated into windows.

However, the voyage of perovskite solar cells is not without difficulties. Their long-term stability and endurance in real-world environments are currently being scrutinized. Researchers are working ceaselessly to address deterioration concerns, notably those caused by moisture and UV radiation exposure. If these obstacles can be overcome, perovskite solar cells might revolutionize the solar industry by providing a cheaper and more efficient alternative to standard silicon-based panels.

Organic Photovoltaics (OPVs)

Organic photovoltaics (OPVs) are another interesting new discovery in solar technology. These cells are constructed of carbon-based materials and may be printed on flexible substrates, making them lightweight, portable, and potentially low-cost to manufacture. OPVs are suited for applications where standard solar panels are unfeasible, such as wearable technology or embedded in ordinary things. While OPVs are currently less efficient than other technologies, their low-cost manufacture and adaptability position them as a crucial participant in the future of solar energy. As research advances, we may anticipate major increases in their performance and durability, bringing them closer to commercial feasibility.

Solar Energy Storage Innovations

As solar energy becomes more widely used, the necessity for effective energy storage options grows. Solar power's intermittent nature—available only when the sun shines—requires the development of innovative storage technology to assure a reliable energy supply.

The Rise of Solid-State Batteries

Solid-state batteries represent a significant advancement in energy storage technology. Solid-state batteries, as opposed to standard lithium-ion batteries, employ a solid electrolyte, which has various advantages, including being safer, having a better energy density, and charging faster. These batteries might play an important role in more effectively storing solar energy, particularly in applications where space and safety are valued.

Solid-state batteries are still in the development phases, with issues such as high production costs and scalability to address. However, as technology advances, we could soon see these batteries powering everything from electric automobiles to large-scale solar energy storage systems.

Hybrid Solar-Wind Systems

The future of renewable energy will most likely require a combination of many technologies working together to produce a regular and dependable energy source. Hybrid solar-wind systems are a prime example of this method. These systems, which combine solar panels and wind turbines, can generate power continually regardless of the weather.

This hybrid method not only increases energy output but also eliminates the need for large-scale energy storage systems, making it a more cost-effective and efficient way to use renewable energy. As technology progresses, we should expect greater widespread usage of hybrid systems, especially in countries with changeable weather patterns.

Solar in Urban Planning

As metropolitan areas expand, incorporating solar technology into the fabric of our cities will be critical to accomplishing sustainability objectives. The future of solar power is not just about generating energy, but also about how we integrate it into our daily lives and constructed surroundings.

Solar-Integrated Infrastructure

One of the most interesting advances in urban solar technology is the incorporation of solar panels into infrastructure. Imagine roadways, sidewalks, and parking lots outfitted with solar panels that generate power while enduring the wear and tear of regular use. While yet at the experimental stage, solar-integrated infrastructure has the potential to turn cities into massive energy-generating networks, decreasing dependency on centralized power plants.

Solar Skins and Building-Integrated Photovoltaic (BIPV)

Building-Integrated Photo voltaic (BIPV) is another significant advancement in urban solar technology. Unlike standard solar panels, which are put on top of existing buildings, BIPV is completely incorporated into the building components. This implies that windows, facades, and roofs may all be constructed to collect solar energy while maintaining aesthetics.

Solar skins, a subset of BIPV, go a step further by providing customizable designs that can resemble typical building materials such as brick and wood. This enables architects and designers to put solar electricity into buildings without changing their look, making it simpler to integrate renewable energy into urban contexts.

Solar Power and Electric Vehicles (EVs)

The combination of solar electricity with electric vehicles (EVs) is a game changer in the search for environmentally friendly transportation. Imagine driving an automobile powered entirely by the sun, eliminating the need for fossil fuels and lowering your carbon impact to nearly zero. This is no longer a faraway fantasy, but rather a fast-approaching reality.

Solar panels are getting more efficient and inexpensive, allowing them to be integrated directly into automobiles. Some pioneering businesses are already creating automobiles with solar roofs, which can capture sunshine to increase driving range. This connection makes EVs more eco-friendly and independent from the grid, allowing users to charge their cars anywhere the sun shines.

Solar Charging Stations

As the number of electric vehicles on the road grows, so will the demand for charging infrastructure. Solar-powered charging stations provide a sustainable response to this need. These stations, which frequently feature big solar arrays, provide clean, renewable energy for charging EVs. They may be deployed in a variety of sites, from cities to distant areas, ensuring that clean energy is available to everyone. Furthermore, solar charging stations may function independently of the standard electrical grid. This makes them perfect for off-grid regions or during power outages, delivering a consistent source of energy when it's most required.

Solar-Powered Electric Cars

Solar-powered electric vehicles are the ultimate of sustainable mobility. These automobiles may capture sunshine to create electricity by embedding solar panels directly into the vehicle's body, either augmenting or totally charging the car's battery. While present technology restricts the amount of electricity that can be generated in this manner, continued advances in solar efficiency and battery storage promise to make entirely solar-powered automobiles a feasible choice in the near future.

These advances aim not just to reduce emissions, but also to build a more robust and autonomous transportation system. Solar-powered automobiles might transform how we think about energy use, allowing you to drive without ever having to tap into the grid.

Decentralized Solar Networks

One of the most intriguing advancements in solar technology is the proliferation of decentralized solar networks. These networks let people and groups to create and exchange solar electricity, lowering dependency on centralized power grids and enhancing energy resilience.

In a decentralized solar network, homeowners and businesses with solar panels can generate extra electricity and feed it into a local microgrid. This energy may then be distributed to others in the network, resulting in a more efficient and sustainable energy system. Decentralized networks also enable communities to govern their energy demands, lowering costs and enhancing energy security.

Community Solar Projects

Community solar projects are an effective strategy to increase public access to renewable energy. These projects enable several families or businesses to get the benefits of a single solar system. Participants can buy or lease a section of the solar array and receive credits on their electricity bills for the energy produced. This concept is especially useful for people who are unable to install solar panels on their own land, such as renters or those with shaded roofs. Community solar projects also encourage local investment in renewable energy, which keeps money inside the community and fosters a feeling of communal responsibility for sustainability.

Peer-to-Peer Energy Trading

Peer-to-peer (P2P) energy trading is a novel idea in which individuals purchase and sell power directly to one another, bypassing traditional utility corporations. Solar-powered P2P networks allow homeowners with solar panels to sell extra energy to neighbors, resulting in a localized energy market.

This technique not only provides financial incentives for producing renewable energy, but it also increases energy independence. With the proper technology, P2P energy trading may make energy markets more competitive, lowering prices and making sustainable energy more accessible to all.

Space-Based Solar Power

Looking beyond our globe, space-based solar power is a bold idea that might transform the way we create energy. By installing solar panels in orbit, where sunlight is continuous and unhindered by the atmosphere, we might theoretically capture massive amounts of energy and beam it back to Earth.

Harvesting Solar Energy in Space

The concept of gathering solar energy in orbit is not new, but technological improvements are bringing it closer to reality. Space-based solar arrays might capture solar energy around the clock, regardless of weather or nightfall. This energy would then be transferred to Earth by microwave or laser beams, where it could be transformed back into electricity and put into the system.

The concept has substantial technological and financial obstacles, but the potential advantages are huge. Space-based solar power might provide a nearly infinite source of renewable energy, helping to fulfill the world's expanding energy demands while without contributing to climate change.

The Potential of Space Solar Satellites

Space solar satellites hold the key to realizing the potential of space-based solar electricity. These spacecraft, outfitted with gigantic solar arrays, would circle the Earth and constantly harvest solar energy. The energy would be sent back to Earth by modern technologies, such as wireless power transfer.

Although currently in the experimental stage, space solar satellites have the potential to become a key component of our global energy system in the future. They provide a technique to create clean, renewable energy on a scale that is unattainable with ground-based solar panels alone.

Artificial Intelligence and Solar Energy

AI is increasingly altering sectors, including solar energy. AI integration into solar technology is no more a pipe dream; it is already transforming how we create, store, and manages solar electricity.

AI for Solar Panel Maintenance and Monitoring

Solar panel maintenance has historically been a labor-intensive operation that requires frequent inspections and manual adjustments. However, artificial intelligence (AI) is altering the game. Solar panels may now be monitored in real time using AI-powered devices, detecting faults before they become important. AI algorithms evaluate data from sensors put on solar panels to detect performance reductions, probable problems, and even predict maintenance needs.

Consider a situation in which your solar panels alert you to a possible problem before you observe a decline in energy production. AI can determine which panels require cleaning, maintenance, or replacement, minimizing downtime and guaranteeing peak performance. This proactive strategy not only saves time and money, but it also increases the lifespan of solar installations.

Predictive Analytics for Solar Power Optimization

AI isn't only about solving issues; it's also about improving performance. Predictive analytics enabled by AI can anticipate weather patterns, energy demand, and other variables in order to maximize solar power output. For example, by forecasting cloud cover or other weather occurrences, AI may alter solar panel angles or manage energy storage to maintain continuous power generation. These prediction algorithms also assist to balance supply and demand. In areas where solar power is a major component of the energy mix, AI can estimate energy output and adapt grid operations appropriately, averting blackouts or energy shortages. This degree of accuracy is critical as we progress towards a future where renewable energy sources like solar play a prominent role.

The Role of Government and Policy in Solar Expansion

While technology drives innovation, government and policy play an important role in the growth of solar energy. Governments worldwide are rapidly realizing the value of solar energy and developing measures to boost its use.

Global Policy Initiatives

Globally, legislative actions are being implemented to promote solar energy as an important component of the renewable energy portfolio. From the Paris Agreement to the European Green Deal, international pledges set lofty goals for solar energy adoption. These efforts seek not just to cut carbon emissions, but also to encourage innovation and economic growth in the renewable energy sector.

For example, India's National Solar Mission intends to position the country as a global leader in solar energy, with a goal of reaching 100 GW by 2022. Similarly, the United States' Solar Energy systems office is sponsoring research and development to reduce prices and boost the efficiency of solar systems.

Incentives and Regulations for Solar Adoption

To encourage solar adoption, governments provide a variety of incentives and laws that make solar power more accessible and inexpensive. Individuals and companies installing solar systems can benefit from tax incentives, rebates, and subsidies. Feed-in tariffs and net metering rules enable solar energy providers to sell excess electricity back to the grid, giving them an additional financial incentive.

Regulations also have an important impact. Building rules are being revised to mandate solar panels in all new structures. Zoning restrictions are being changed to allow for large-scale solar farms. These regulations are critical for providing a favorable climate for solar energy expansion.

Challenges and Solutions for Future Solar Adoption

As bright as solar technology's future appears, there are hurdles that must be overcome in order for it to be widely used. These issues vary from coping with solar waste to addressing the intermittent nature of solar electricity output.

Addressing Solar Waste and Recycling

Solar waste is one of the solar industry's least-discussed challenges. Solar panels have a lifespan of roughly 25-30 years, after which they must be replaced. As the first generations of solar panels near the end of their life cycle, the subject of what to do with them becomes urgent.

The problem is in the recycling process. Solar panels are comprised of complicated components such as glass, silicon, and metals, necessitating sophisticated recycling methods. Unfortunately, recycling facilities are scarce, and recycling is sometimes more expensive than disposal. However, novel solutions are developing. Some businesses are designing panels that are easier to disassemble and recycle, while others are researching more effective recycling technologies. Governments are now beginning to enact policies that require recycling and promote the development of recycling infrastructure.

Overcoming Intermittency Issues with Solar Power

One of the most significant issues of solar power is its intermittency—the fact that solar energy output is not continuous and is dependent on sunshine availability. Solar panels cannot generate power constantly on cloudy days or at night, which can be problematic for guaranteeing a consistent energy supply.

To address this issue, energy storage technologies such as batteries are becoming more significant. By storing surplus energy created during sunny hours, these devices can offer electricity when the sun is not shining. In addition, AI and smart grid technologies are being developed to better forecast and regulate the flow of solar energy, ensuring that it is used efficiently and effectively throughout the system.

Another viable alternative is to create hybrid systems that combine solar and other renewable energy sources, such as wind or hydro, to give a more regular power supply. These systems may balance the advantages and disadvantages of each energy source, ensuring that energy is always accessible when required.

www.ingramcontent.com/pod-product-compliance
Lightning Source LLC
Chambersburg PA
CBHW052137220526
45471CB00004B/1422